Soil and Water Chemistry: Principles, Properties and Applications

Soil and Water Chemistry: Principles, Properties and Applications

Dr. Prashant Kumar Sirohi

RANDOM

RANDOM PUBLICATIONS

NEW DELHI (INDIA)

Soil and Water Chemistry: Principles, Properties and Applications

ISBN 978-93-5111-589-2

Published in 2015 in India by

RANDOM PUBLICATIONS

4376-A/4B, Gali Murari Lal, Ansari Road
New Delhi-110 002
Phone : +9111-43580356, 011-23289044, 011-43142548
e-mail: sales@randompublications.com,
info@randompublications.com, randomexports@gmail.com

Reprinted 2026

Type Setting by : Friends Media, Delhi-110089

Preface

Soil chemistry is the study of the chemical characteristics of soil. Soil chemistry is affected by mineral composition, organic matter and environmental factors. Until the late 1960s, soil chemistry focused primarily on chemical reactions in the soil that contribute to pedogenesis or that affect plant growth. Since then, concerns have grown about environmental pollution, organic and inorganic soil contamination and potential ecological health and environmental health risks. Consequently, the emphasis in soil chemistry has shifted from pedology and agricultural soil science to an emphasis on environmental soil science.

Water plays an important role as a chemical substance. Its many important functions include being a good solvent for dissolving many solids, serving as an excellent coolant both mechanically and biologically, and acting as a reactant in many chemical reactions. Blood, sweat and tears... all solutions of water. As chemists we consider water from many perspectives. It is our role to use physical and mathematical laws in application for useful purposes, including diverse perspectives such as living systems, materials and energy. The world of the chemist is a small world - atomic, molecular - which plays a large part in making our lives healthy, comfortable, and hopeful. Because of the diversity of the chemical world, it would be difficult to touch upon all of the applications of water. And for the same reason, it would be impossible to discuss the chemical aspects of water without touching upon the physical, mathematical, and biological aspects of the subject.

I would like to thank my team for standing beside me throughout my career and writing this book. My special thanks go to" Random Publications" who have published the book.

– Dr. Prashant Kumar Sirohi

Contents

1

Soil Description and Classification

It is necessary to adopt a formal system of soil description and classification in order to describe the various materials found in ground investigation. Such a system must be comprehensive (covering all but the rarest of deposits), meaningful in an engineering context *(so that engineers will be able to understand and interpret)* and yet relatively concise. It is important to distinguish between description and classification: Description of soil is a statement describing the physical nature and state of the soil.

It can be a description of a sample, or a soil *in situ*. It is arrived at using visual examination, simple tests, observation of site conditions, geological history, etc. Soil classification is the separation of soil into classes or groups each having similar characteristics and potentially similar behaviour. A classification for engineering purposes should be based mainly on mechanical properties, *e.g.* permeability, stiffness, strength. The class to which a soil belongs can be used in its description.

BASIC CHARACTERISTICS OF SOILS

Soils consist of grains (mineral grains, rock fragments, etc.) with water and air in the voids between grains. The water and air contents are readily changed by changes in conditions and location: soils can be perfectly dry (have no water content) or be fully saturated (have no air content) or be partly saturated (with both air and water present).

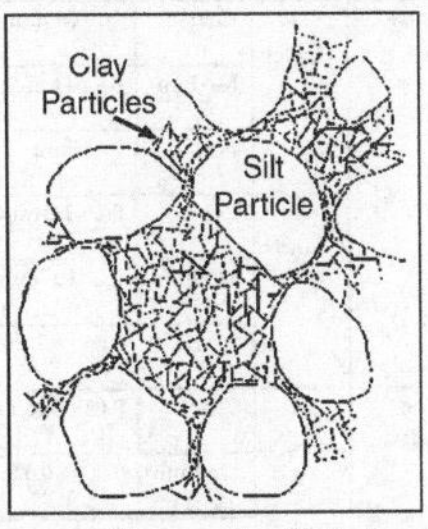

Although the size and shape of the solid (granular) content rarely changes at a given point, they can vary considerably from point to point.

First of all, consider soil as a engineering material-it is not a coherent solid material like steel and concrete, but is a particulate material. It is important to understand the significance of particle size, shape and composition, and of a soil's internal structure or fabric.

Soil as an Engineering Material

The term "soil" means different things to different people: To a geologist it represents the products of past surface processes. To a pedologist it represents currently occurring physical and chemical processes.

To an engineer it is a material that can be: built on: foundations to buildings, bridges. built in: tunnels, culverts, basements. built with: roads, runways, embankments, dams. supported: retaining walls, quays. Soils may be described in different ways by different people for their different purposes. Engineers' descriptions give engineering terms that will convey some sense of a soil's current state and probable susceptibility to future changes (*e.g.* in loading, drainage, structure, surface level). Engineers are primarily interested in a soil's mechanical properties: strength, stiffness, permeability. These depend primarily on the nature of the soil grains, the current stress, the water content and unit weight.

Size Range of Grains

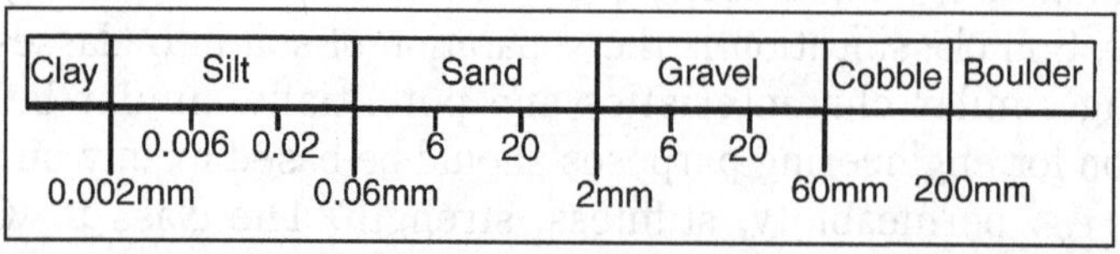

The range of particle sizes encountered in soil is very large: from boulders with a controlling dimension of over 200mm down to clay particles less than 0.002mm (2μm). Some clays contain particles less than 1 mm in size which behave as colloids, *i.e.* do not settle in water due solely to gravity.

Very coarse soils	Boulders		> 200 mm
	Cobbles		60 - 200 mm
Coarse soils	G Gravel	Coarse	20 - 60 mm
		Medium	6 - 20 mm
		Fine	2 - 6 mm
	S Sand	Coarse	0.6 - 2.0 mm
		Medium	0.2 - 0.6 mm
		Fine	0.06 - 0.2 mm
Fine soils	M Silt	Coarse	0.02 - 0.06 mm
		Medium	0.006 - 0.02 mm
		Fine	0.002 - 0.006 mm
	C Clay		< 0.002 mm

In the British Soil Classification System, soils are classified into named Basic Soil Type groups according to size, and the groups further divided into coarse, medium and fine sub-groups:

Aids to Size Identification

Soils possess a number of physical characteristics which can be used as aids to size identification in the field. A handful of soil rubbed through the fingers can yield the following:

Sand (and coarser) particles are visible to the naked eye. SILT particles become dusty when dry and are easily brushed off hands and boots. Clay particles are greasy and sticky when wet and hard when dry, and have to be scraped or washed off hands and boots.

SHAPE OF GRAINS

- Shape characteristics of Sands grains
- Shape characteristics of Clay grains
- Specific surface

The majority of soils may be regarded as either SANDS or Clay:

Sands include gravelly sands and gravel-sands. Sand grains are generally broken rock particles that have been formed by physical weathering, or they are the resistant components of rocks broken down by chemical weathering. Sand grains generally have a rotund shape.

Clays include silty clays and clay-silts; there are few pure silts (*e.g.* areas formed by windblown Löess). Clay grains are usually the product of chemical weathering or rocks and soils. Clay particles have a flaky shape.

THERE ARE MAJOR DIFFERENCES INCHARACTERISTICS OF SAND GRAINS

Sand and larger-sized grains are rotund. Coarse soil grains (silt-sized, sand-sized and larger) have different shape characteristics and surface roughness depending on the amount of wear during transportation (by water, wind or ice), or after crushing in manufactured aggregates. They have a relatively low specific surface (surface area).

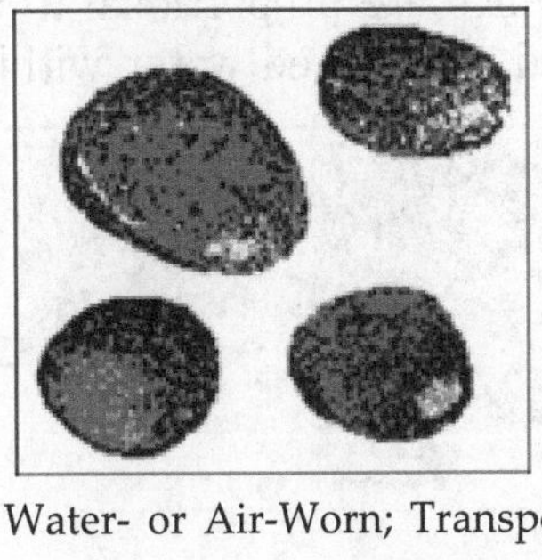

Fig. Rounded: Water- or Air-Worn; Transported Sediments

Fig. Irregular: Irregular Shape With Round Edges; Glacial Sediments (Sometimes Sub-Divided into 'Sub-Rounded' and 'Sub-angular')

Fig. Angular: Flat Faces and Sharp Edges; Residual Soils, Grits

Fig. Flaky: Thickness Small Compared to Length/Breadth; Clays

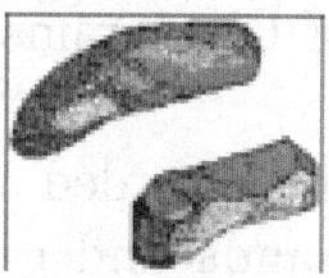

Fig. Elongated: Length Larger Than Breadth/Thickness; Scree, Broken Flagstone

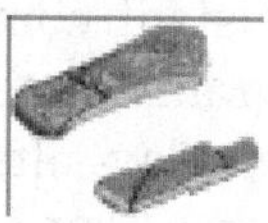

Fig. Flaky and Elongated: Length>Breadth>Thickness; Broken Schists and Slates

SHAPE CHARACTERISTICS OF CLAY GRAINS

CLAY particles are flaky. Their thickness is very small relative to their length and breadth, in some cases as thin as 1/100th of the length. They therefore have high to very high specific surface values. These surfaces carry a small negative electrical charge, that will attract the positive end of water molecules. This charge depends on the soil mineral and may be affected by an electrolite in the pore water. This causes some additional forces between the soil grains which are proportional to the specific surface. Thus a lot of water may be held asadsorbed water within a clay mass.

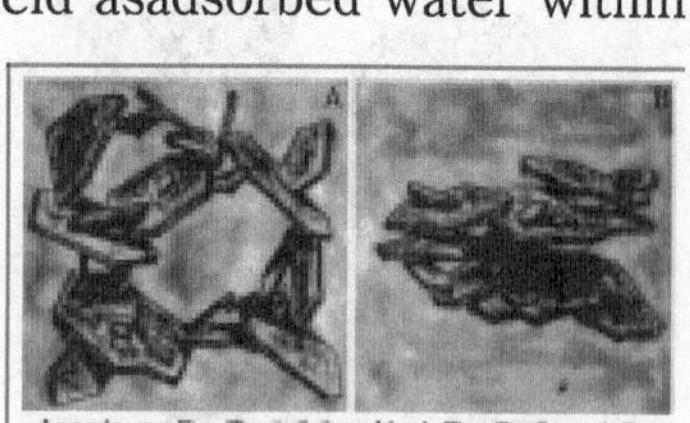

Anordnung Der Tonteilchen Nach Der Sedimentation
A) in Meerwasser and B) in Süsswasser

SPECIFIC SURFACE

Examples

Specific surface is the ratio of surface area per unit wight. Surface forces are proportional to surface area (*i.e.* to d^2).

Hence, specific surface is a measure of the relative contributions of surface forces and self-weight forces. The specific surface of a 1mm cube of quartz (ρ = 2.65gm/cm^3) is 0.00023 m^2/N Sand grains (size 2.0–0.06mm) are close to cubes or spheres in shape, and have specific surfaces near the minimum value. Clay particles are flaky and have much greater specific surface values.

Examples of Specific Surface

The more elongated or flaky a particle is the greater will be its specific surface.

Cubes

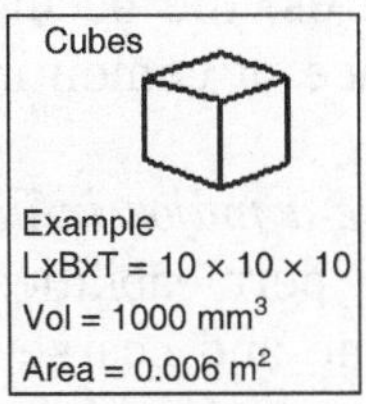

Rods

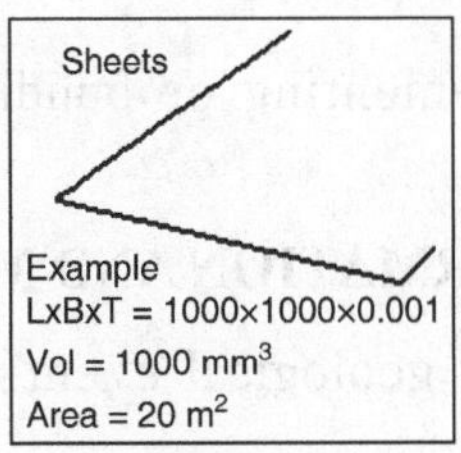

Fig. 1.11

Sheets

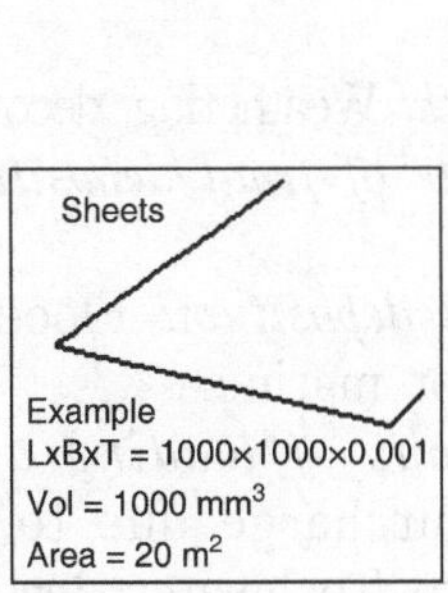

Table. Examples of Mineral Grain Specific Surfaces

Mineral/Soil	Grainwidth d (mm)	Thickness	Specific Surface m^2/N
Quartz grain	100	d	0.0023
Quartz sand	2.0–0.06	»d	0.0001–0.004
Kaolinite	2.0–0.3	»0.2d	2
Illite	2.0–0.2	»0.1d	8
Montmorillonite	1.0–0.01	»0.01d	80

STRUCTURE OR FABRIC

Natural soils are rarely the same from one point in the ground to another.

The content and nature of grains varies, but more importantly, so does the arrangement of these. The arrangement and organisation of particles and other features within a soil mass is termed its structure or fabric.

This includes bedding orientation, stratification, layer thickness, the occurrence of joints and fissures, the occurrence of voids, artefacts, tree roots and nodules, the presence of cementing or bonding agents between grains.

Structural features can have a major influence on in situ properties:

- Vertical and horizontal permeabilities will be different in alternating layers of fine and coarse soils.
- The presence of fissures affects some aspects of strength.
- The presence of layers or lenses of different stiffness can affect stability.
- The presence of cementing or bonding influences strength and stiffness.

ORIGINS, FORMATION AND MINERALOGY

Soils are the results of geological events (except for the very small amount produced by man).

The nature and structure of a given soil depends on the geological processes that formed it:

- *Breakdown of parent rock:* Weathering, decomposition, erosion.
- *Transportation to site of final deposition:* Gravity, flowing water, ice, wind.
- *Environment of final deposition:* Flood plain, river terrace, glacial moraine, lacustrine or marine.
- *Subsequent conditions of loading and drainage:* Little or no surcharge, heavy surcharge due to ice or overlying deposits, change from saline to freshwater, leaching, contamination.

ORIGINS OF SOILS FROM ROCKS

All soils originate, directly or indirectly, from solid rocks in the Earth's crust:

- *Igneous Rocks*: Crystalline bodies of cooled magma *e.g.* granite, basalt, dolerite, gabbro, syenite, porphyry
- *Sedimentary Rocks*: Layers of consolidated andcemented sediments, mostly formed in bodies of water (seas, lakes, etc.) *e.g.* limestone, sandstones, mudstone, shale, conglomerte
- *Metamorphic Rocks*: Formed by the alteration of existing rocks due to heat from igneous intrusions (*e.g.* marble, quartzite, hornfels) or pressure due to crustal movement (*e.g.* slate, schist, gneiss).

WEATHERING OF ROCKS

Physical Weathering

Physical or mechanical processes taking place on the Earth's surface, including the actions of water, frost, temperature changes, wind and ice; cause disintegration and wearing. The products are mainly coarse soils (silts, sands and gravels). Physical weathering produces Very Coarse soils and Gravels consisting of broken rock particles, but Sands and Silts will be mainly consists of mineral grains.

Chemical Weathering

Chemical weathering occurs in wet and warm conditions and consists of degradation by decomposition and/or alteration. The results of chemical weathering are generally fine soils with separate mineral grains, such as Clays and Clay-Silts. The type of clay mineral depends on the parent rock and on local drainage. Some minerals, such as quartz, are resistant to the chemical weathering and remain unchanged.

Quartz

A resistant and enduring mineral found in many rocks (*e.g.* granite, sandstone). It is the principal constituent of sands and silts, and the most abundant soil mineral. It occurs as equidimensional hard grains.

Haematite

A red iron (ferric) oxide: resistant to change, results from extreme weathering. It is responsible for the widespread red or pink colouration in rocks and soils. It can form a cement in rocks, or a duricrust in soils in arid climates.

Micas

Flaky minerals present in many igneous rocks. Some are resistant, *e.g.* muscovite; some are broken down, *e.g.* biotite.

Clay Minerals

These result mainly from the breakdown of feldspar minerals. They are very flaky and therefore have very large surface areas. They are major constituents of clay soils, although clay *soil* also contains silt sized particles.

CLAY MINERALS

Clay minerals are produced mainly from the chemical weathering and decomposition of feldspars, such as orthoclase and plagioclase, and some micas.

They are small in size and very flaky in shape. The key to some of the properties of clay soils, *e.g.* plasticity, compressibility, swelling/shrinkage potential, lies in the structure of clay minerals.

There are three main groups of clay minerals:

1. *Kaolinites:* (Include kaolinite, dickite and nacrite) formed by the decomposition of orthoclase feldspar (*e.g.* in granite); kaolin is the principal constituent in china clay and ballclay.
2. *Illites:* (Include illite and glauconite) are the commonest clay minerals; formed by the decomposition of some micas and feldspars; predominant in marine clays and shales.
3. *Montmorillonites:* (Also called smectites or fullers' earth minerals) (include calcium and sodium momt-morillonites, bentonite and vermiculite) formed by the alteration of basic igneous rocks containing silicates rich in Ca and Mg; weak linkage by cations (*e.g.* Na^{+}, Ca^{++}) results in high swelling/shrinking potential

TRANSPORTATION AND DEPOSITION

The effects of weathering and transportation largely determine the basic nature of the soil (*i.e.* the size, shape, composition and distribution of the grains). The environment into which deposition takes place, and subsequent geological events that take place there, largely determine the state of the soil, (*i.e.* density, moisture content) and the structure or fabric of the soil (*i.e.* bedding, stratification, occurrence of joints or fissures, tree roots, voids, etc.)

Transportation

Due to combinations of gravity, flowing water or air, and moving ice. In water or air: grains become sub-rounded or rounded, grain sizes are sorted, producing poorly-graded deposits. *In moving ice*: grinding and crushing occur, size distribution

becomes wider, deposits are well-graded, ranging from rock flour to boulders.

Deposition

In flowing water, larger particles are deposited as velocity drops, *e.g.* gravels in river terraces, sands in floodplains and estuaries, silts and clays in lakes and seas.

In still water: horizontal layers of successive sediments are formed, which may change with time, even seasonally or daily.

- *Deltaic and shelf deposits:* Often vary both horizontally and vertically.
- From glaciers, deposition varies from well-graded basal tills and boulder clays to poorly-graded deposits in moraines and outwash fans.
- *In arid conditions:* Scree material is usually poorly-graded and lies on slopes.
- Wind-blown Loess is generally uniformly-graded and false-bedded.

LOADING AND DRAINAGE HISTORY

The current state (*i.e.* density and consistency) of a soil will have been profoundly influenced by the history of loading and unloading since it was deposited. Changes in drainage conditions may also have occurred which may have brought about changes in water content.

Loading/Unloading History

Initial Loading

During deposition the load applied to a layer of soil increases as more layers are deposited over it; thus, it is compressed and water is squeezed out; as deposition continues, the soil becomes stiffer and stronger.

Unloading

The principal natural mechanism of unloading is erosion of overlying layers. Unloading can also occur as overlying ice-sheets and glaciers retreat, or due to large excavations made by man.

Soil expands when it is unloaded, but not as much as it was initially compressed; thus it stays compressed-and is said to be overconsolidated. The degree of overconsolidation depends on the history of loading and unloading.

Drainage History

Chemical Changes

Some soils initially deposited loosely in saline water and then inundated with fresh water develop weak collapsing structure. In arid climates with intermittent rainy periods, cycles of wetting and drying can bring minerals to the surface to form a cemented soil.

Climate Changes

Some clays (*e.g.* montmorillonite clays) are prone to large volume changes due to wetting and drying; thus, seasonal changes in surface level occur, often causing foundation damage, especially after exceptionally dry summers. Trees extract water from soil in the process of evapotranspiration; The soil near to trees can therefore either shrink as trees grow larger, or expand following the removal of large trees.

GRADING AND COMPOSITION

The recommended standard for soil classification is the British Soil Classification System, and this is detailed in BS 5930 Site Investigation.

COARSE SOILS

- Particle size tests
- Typical grading curves
- Grading characteristics
- Sieve analysis example

Table. Coarse Soils Are Classified Principally on The Basis of Particle Size and Grading.

<table>
<tr><td rowspan="2">Very coarse soils</td><td colspan="2">Boulders</td><td>> 200 mm</td></tr>
<tr><td colspan="2">Cobbles</td><td>60 - 200 mm</td></tr>
<tr><td rowspan="6">Coarse soils</td><td rowspan="3">G
Gravel</td><td>Coarse</td><td>20 - 60 mm</td></tr>
<tr><td>Medium</td><td>6 - 20 mm</td></tr>
<tr><td>Fine</td><td>2 - 6 mm</td></tr>
<tr><td rowspan="3">S
Sand</td><td>Coarse</td><td>0.6 - 2.0 mm</td></tr>
<tr><td>Medium</td><td>0.2 - 0.6 mm</td></tr>
<tr><td>Fine</td><td>0.06 - 0.2 mm</td></tr>
</table>

Particle Size Tests

The aim is to measure the distribution of particle sizes in the sample. When a wide range of sizes is present, the sample will be sub-divided, and separate tests carried out on each sub-sample. Full details of tests are given in BS 1377: "Methods of test for soil for civil engineering purposes".

Particle-Size Tests

- *Wet sieving:* To separate fine grains from coarse grains is carried out by washing the soil specimen on a 60mm sieve mesh.
- *Dry sieving:* Analyses can only be carried out on particles > 60 mm. Samples (with fines removed) are dried and shaken through a nest of sieves of descending size.
- *Sedimentation:* Is used only for fine soils. Soil particles are allowed to settle from a suspension. The decreasing density of the suspension is measured at time intervals. Sizes are determined from the settling velocity and times recorded. Percentages between sizes are determined from density differences.

Particle-Size Analysis

The cumulative percentage quantities finer than certain sizes (*e.g.* passing a given size sieve mesh) are determined by weighing. Points are then plotted of per cent finer (passing) against log size. A smooth S-shaped curve drawn through these points is called a grading curve. The position and shape of the grading curve determines the soil class. Geometrical grading characteristics can be determined also from the grading curve.

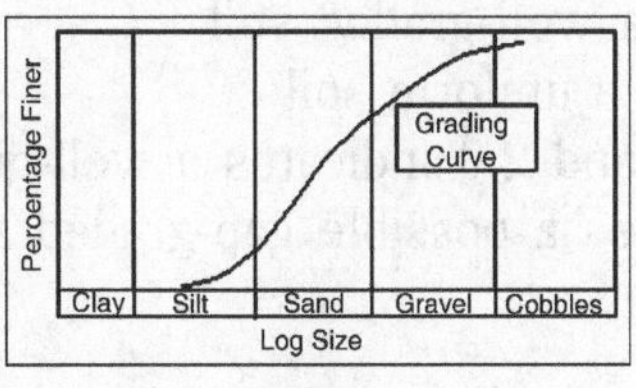

Typical Grading Curves

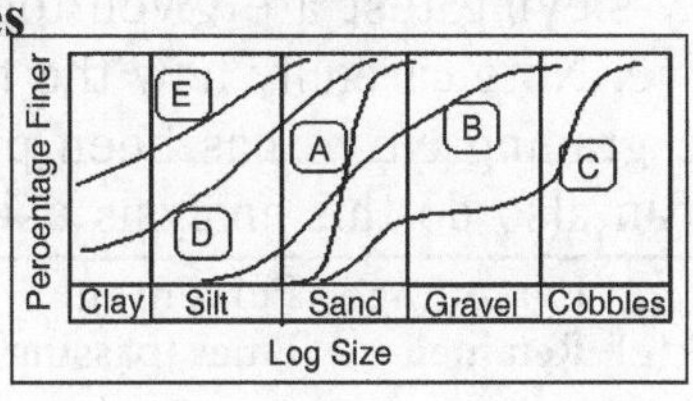

Both the position and the shape of the grading curve for a soil can aid its identity and description.

Some typical grading curves are shown in the figure 1.16:

- A–a poorly-graded medium Sand (probably estuarine or flood-plain alluvium)
- B–a well-graded Gravel-Sand (*i.e.* equal amounts of gravel and sand)
- C–a gap-graded Cobbles-Sand
- D–a sandy Silt (perhaps a deltaic or estuarine silt)
- E–a typical silty Clay

Grading Characteristics

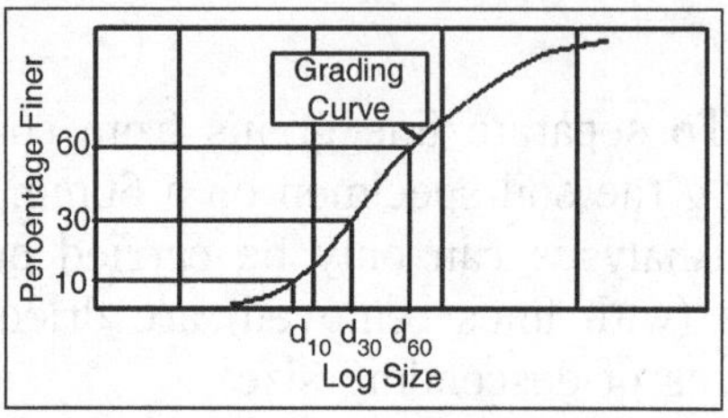

A grading curve is a useful aid to soil description. Grading curves are often included in ground investigation reports. Results of grading tests can be tabulated using geometric properties of the grading curve. These properties are called grading characteristics

First of all, three points are located on the grading curve:

1. d_{10} = The maximum size of the smallest 10% of the sample
2. d_{30} = The maximum size of the smallest 30% of the sample
3. d_{60} = The maximum size of the smallest 60% of the sample

From these the grading characteristics are calculated:

- *Effective size:* d_{10}
- *Uniformity coefficient:* $C_u = d_{60}/ d_{10}$
- *Coefficient of gradation:* $C_k = d_{30}^2/ d_{60}\, d_{10}$

Both C_u and C_k will be 1 for a single-sized soil:

- $C_u > 5$ indicates a well-graded soil
- $C_u < 3$ indicates a uniform soil
- C_k between 0.5 and 2.0 indicates a well-graded soil
- $C_k < 0.1$ indicates a possible gap-graded soil

Sieve Analysis Example

The results of a dry-sieving test are given together with the grading analysis and grading curve. Note carefully how the tabulated results are set out and calculated. The grading curve has been plotted on special semi-logarithmic paper; you can also do this analysis using a spreadsheet.

Sieve Mesh **Size (mm)**	Mass **Retained (g)**	Percentage **Retained**	Percentage **Finer (passing)**

14.0	0	0	100.0
10.0	3.5	1.2	98.8
6.3	7.6	2.6	86.2
5.0	7.0	2.4	93.8
3.35	14.3	4.9	88.9
2.0	21.1	7.2	81.7
1.18	56.7	19.4	62.3
0.600	73.4	25.1	37.2
0.425	22.2	7.6	29.6
0.300	26.9	9.2	20.4
0.212	18.4	6.3	14.1
0.150	15.2	5.2	8.9
0.063	17.5	6.0	2.9
Pan	8.5	2.9	
Total	292.3	100.0	

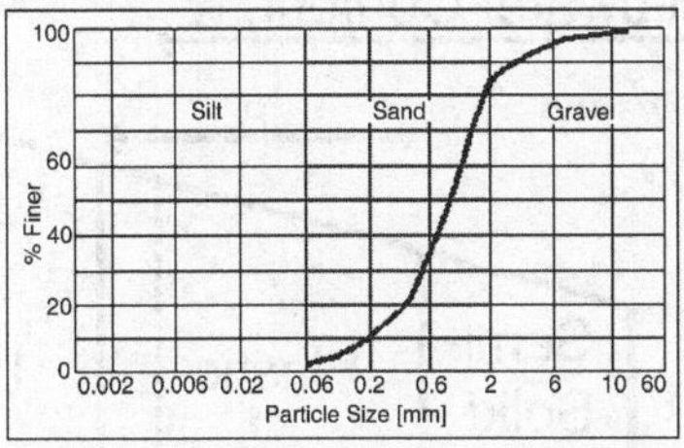

The soil comprises: 18% gravel, 45% coarse sand, 24% medium sand, 10% fine sand, 3% silt, and is classified therefore as: a well-graded gravelly Sand

FINE SOILS

- Consistency limits and plasticity
- Plasticity index
- The plasticity chart and classification
- Activity

In the case of fine soils (*e.g.* Clays and Silts), it is the *shape* of the particles rather than their size that has the greater influence on engineering properties. Clay soils have flaky particles to which water adheres, thus imparting the property of plasticity.

Consistency Limits and Plasticity

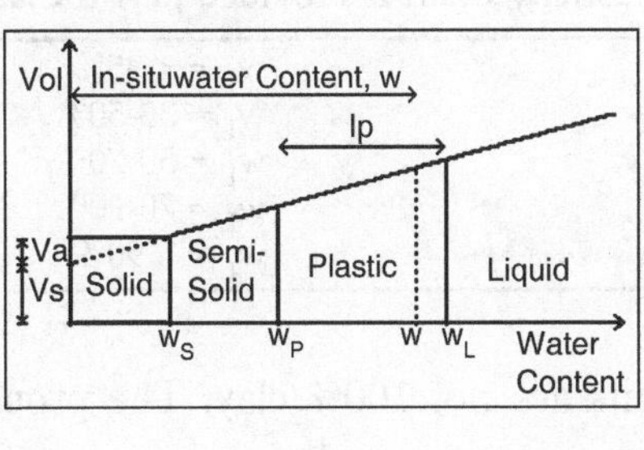

Consistency varies with the water content of the soil. The consistency of a soil can range from (dry) *solid* to *semi-solid* to *plastic* to *liquid* (wet). The water contents at which the consistency changes from one state to the next are called consistency limits (or Atterberg limits).

Two of these are utilised in the classification of fine soils:

1. *Liquid Limit* (w_L)*:* Change of consistency from plastic to liquid
2. *Plastic Limit* (w_P)*:* Change of consistency from brittle/crumbly to plastic

Measures of liquid and plastic limit values can be obtained from laboratory tests.

Plasticity Index

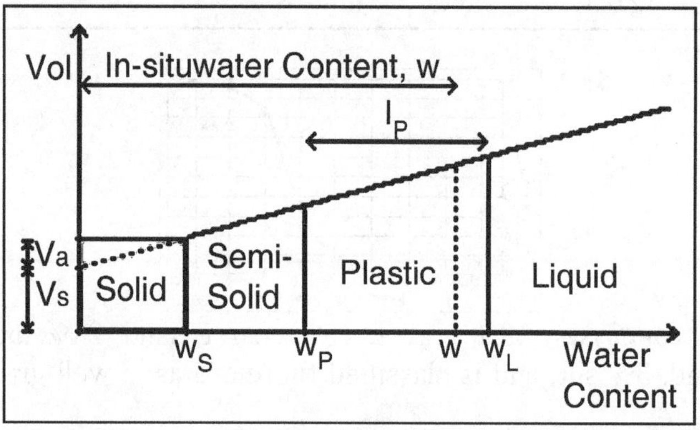

The consistency of most soils in the ground will be plastic or semi-solid. Soil strength and stiffness behaviour are related to the range of plastic consistency. The range of water content over which a soil has a plastic consistency is termed the Plasticity Index (I_P or PI).

I_P = Liquid limit–plastic limit

= $w_L - w_P$

The Plasticity Chart and Classification

In the BSCS fine soils are divided into *ten* classes based on their measured plasticity index and liquid limit values: CLAYS are distinguished from SILTS, and five divisions of plasticity are defined:

Table. A Plasticity Chart is Provided to Aid Classification.

Low plasticity	w_L = < 35%
Intermediate plasticity	w_L = 35-50%
High plasticity	w_L = 50-70%
Very high plasticity	w_L = 70-90%
Extremely high plasticity	w_L = > 90%

Activity

So-called 'clay' soils are not 100% clay. The proportion of clay mineral flakes (< 2 μm size) in a fine soil affects its current state, particularly its tendency to swell and shrink with changes in water content. The degree of plasticity related to the clay content is called the activity of the soil.

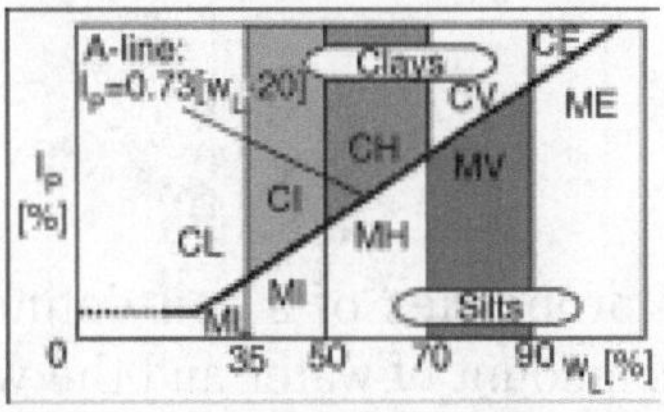

Activity = I_P/ (per cent clay particles) Some typical values are:

Mineral	Activity	Soil	Activity
Muscovite	**0.25**	**Kaolin clay**	**0.4-0.5**
Kaolinite	**0.40**	**Glacial clay and loess**	**0.5-0.75**
Illite	**0.90**	**Most British clays**	**0.75-1.25**
Montmorillonite	**> 1.25**	**Organic estuarine clay**	**> 1.25**

SPECIFIC GRAVITY

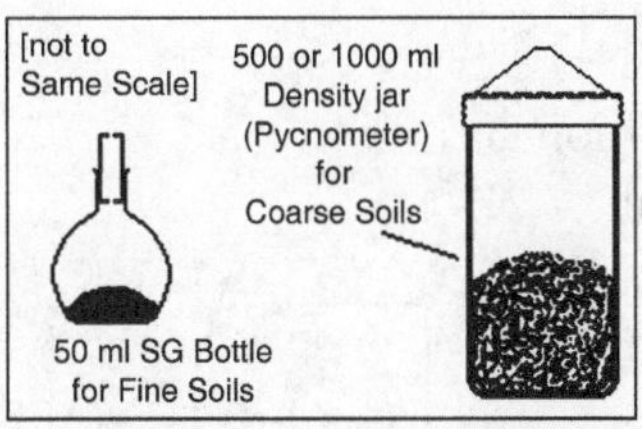

Specific gravity (G_s) is a property of the mineral or rock material forming soil grains. It is defined as

$$G_S = \frac{\text{Mass of a soil grain}}{\text{Mass of an equal volume of water}}$$

Method of Measurement

For fine soils a 50 ml density bottle may be used; for coarse soils a 500 ml or 1000 ml jar. The jar is weighed empty (M_1). A quantity of dry soil is placed in the jar and the jar weighed (M_2). The jar is filled with water, air removed by stirring, and weighed again (M_3). The jar is emptied, cleaned and refilled with water-and weighed again (M_4).

$$G_S = \frac{\text{Mass of soil}}{\text{Mass of water displaced by soil}}$$

$$= \frac{M_2 - M_1}{(M_4 - M_1) - (M_3 - M_2)}$$

[The range of G_s for common soils is 2.64 to 2.72]

VOLUME-WEIGHT PROPERTIES

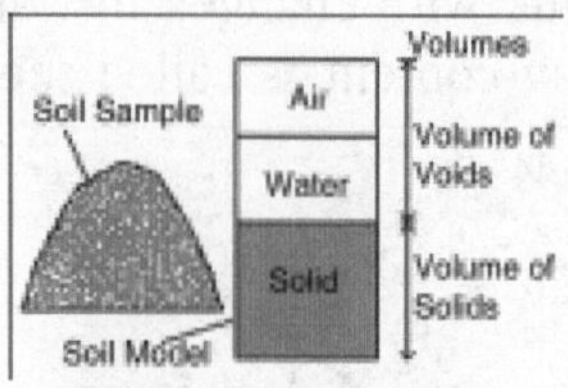

The volume-weight properties of a soil define its state. Measures of the amount of void space, amount of water and the weight of a unit volume of soil are required in engineering analysis and design.

Soil comprises three constituent phases:

1. *Solid:* Rock fragments, mineral grains or flakes, organic matter.
2. *Liquid:* Water, with some dissolved compounds (*e.g.* salts).
3. *Gas:* Air or water vapour.

In natural soils the three phases are intermixed. To aid analysis it is convenient to consider a soil model in which the three phases are seen as separate, but still in their correct proportions.

VOLUMES OF SOLID, WATER AND AIR: THE SOIL MODEL

- Degree of saturation
- Air-voids content

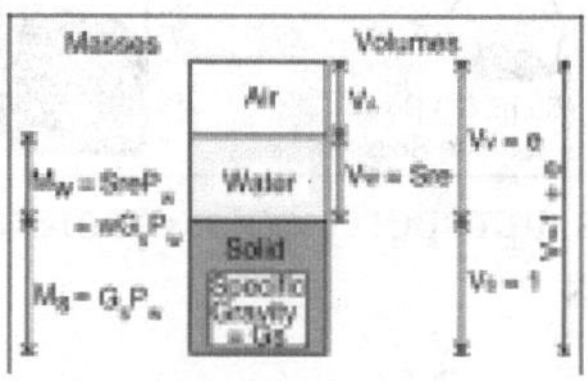

The soil model is given dimensional values for the solid, water and air components: Total volume,

$$V = V_s + V_w + V_a$$

Since the amounts of both water and air are variable, the volume of solids present is taken as the reference quantity.

Thus, the following relational volumetric quantities may be defined:

$$\text{Void ratio}, e = \frac{\text{Volume of voids}}{\text{Volume of solids}} = \frac{V_v}{V}$$

$$\text{Porosity}, n = \frac{\text{Volume of voids}}{\text{Total volume}} = \frac{V_v}{V}$$

$$\text{Specific volume}, v = 1 + e = \frac{V}{V_s}$$

Note also that:

$$n = e/\,(1 + e)$$

$$e = n/\,(1-n)$$

$$v = 1/\,(1-n)$$

Typical void ratios might be 0.3 (*e.g.* for a dense, well graded granular soil) or 1.5 (*e.g.* for a soft clay).

Degree of Saturation

The volume of water in a soil can only vary between zero (*i.e.* a dry soil) and the volume of voids; this can be expressed as a ratio:

$$\text{Degree of saturation}, S_r = \frac{\text{Volume of water}}{\text{Volume of voids}} = \frac{V_w}{V_v}$$

For a perfectly dry soil:

- $S_r = 0$

For a saturated soil:

- $S_r = 1$

Note: In clay soils as the amount water increases the volume and therefore the volume of voids will also increase, and so the degree of saturation may remain at $S_r = 1$ while the actual volume of water is increasing.

Air-Voids Content

The air-voids volume, V_a, is that part of the void space not occupied by water.

$$V_a = V_v - V_w$$
$$= e - e.S_r$$
$$= e.(1 - S_r)$$

Air-voids content, A_v

A_v = (Air-voids volume)/(Total volume)

$= V_a / V$

$= e.(1-S_r)/\,(1+e)$

$= n.(1-S_r)$

For a perfectly *dry* soil:

$$A_v = n$$

For a *saturated* soil:

$$A_v = 0$$

MASSES OF SOLID AND WATER: WATER CONTENT

The mass of air may be ignored. The mass of solid particles is usually expressed in terms of their particle density or grain specific gravity.

Grain Specific Gravity

$$G_s = \frac{\text{Mass of a soil grain}}{\text{Mass of an equal volume of water}}$$

Hence the mass of solid particles in a soil,

$$M_s = V_s.G_s.r_w$$

(r_w = density of water = 1.00Mg/m³)
[Range of G_s for common soils: 2.64–2.72]

Particle Density

r_s = Mass per unit volume of particles
= $G_s.r_w$

The ratio of the mass of water present to the mass of solid particles is called the *water* content, or sometimes the moisture content.

$$\text{Water content, } w = \frac{\text{Mass of water}}{\text{Mass of solids}} = \frac{M_w}{M_s}$$

From the soil model it can be seen that,

$$w = (S_r.e.r_w)/\ (G_s.r_w)$$

Giving the useful relationship:

$$w.G_s = S_r.e$$

DENSITIES AND UNIT WEIGHTS

Density is a measure of the quantity of *mass* in a unit volume of material. Unit weight is a measure of the *weight* of a unit volume of material. There are two basic measures of density or unit weight applied to soils: Dry density is a measure of the amount of *solid* particles per unit volume. Bulk density is a measure of the amount of *solid* + *water* per unit volume.

$$\text{Dry density, } \rho_d = \frac{\text{Mass of solids}}{\text{Total volume}} = \frac{M_s}{V} = \frac{G_s\rho_w}{1+e}$$

$$\text{Bulk density, } \rho = \frac{\text{Total mass}}{\text{Total volume}} = \frac{M_s + M_w}{V} = \frac{G_s\rho_w + S_r e\rho_w}{1+e}$$

The preferred units of density are:
Mg/m³, kg/m³ or g/ml.
The corresponding *unit weights* are:

$$\text{Dry unit weight, } \gamma_d = \frac{\text{Dry weight}}{\text{Total volume}} = \frac{W_s}{V} = \frac{G_s\gamma_w}{1+e} = 9.81\rho d$$

$$\text{Unit weight, } \gamma = \frac{\text{Total weight}}{\text{Total volume}} = \frac{W_s + W_w}{V} = \frac{G_s\gamma_w + S_r e\gamma_w}{1+e} = 9.81\rho$$

Also, it can be shown that,

$$r = r_d(1 + w) \text{ and}$$

$$g = g_d(1 + w)$$

LABORATORY MEASUREMENTS

- Water content
- Unit weight

It is important to quantify the state of a soil immediately it is received in the testing laboratory and just prior to commencing other tests (*e.g.* shear tests, compression tests, etc.).

The water content and unit weight are particularly important, since these could change during transportation and storage. Some physical state properties are calculated following the practical measurement of others; *e.g.* void ratio from porosity, dry unit weight from unit weight and water content.

Water Content

The most usual method of determining the water content of soil is to weigh a small representative specimen, drying it to constant weight and then weighing it again. Drying can be carried out using an electric oven set at 104–105° Celsius or using a microwave oven.

Example: *A sample of soil was placed in a tin container and weighed, after which it was dried in an oven and then weighed again. Calculate the water content of the soil.*

Weight of tin empty	= 16.16 g
Weight of tin + moist soil	= 37.82 g
Weight of tin + dry soil	= 34.68 g
Water content, w	= (Mass of water)/(Mass of dry soil)
	= (37.82–34.68)/ (34.68–16.16)
	= 0.169
Percentage water content	= 16.9%

Unit Weight

- *Clay soils:* Specimens are usually prepared in the form of regular geometric shapes, (*e.g.* prisms, cylinders) of which the volume is easily computed.
- *Sands and gravels:* Specimens have to be placed in a container to determine volume (*e.g.* a cylindrical can).

Example

A soil specimen had a volume of 89.13 ml, a mass before drying of 174.45

g and after drying of 158.73 g; the water content was 9.9%. Determine the bulk and dry densities and unit weights.

Bulk Density

ρ = (Mass of specimen)/(Volume of specimen)
= 174.45/ 89.13 g/ml
= 1.957 Mg/m^3

[1 g/ml = 1 Mg/m^3]

Unit Weight

γ = 9.81m/s^2 × r Mg/m^3
= 19.20 kN/m^3

Dry density

ρ_d = (Mass after drying)/(volume)
= 158.73/ 89.13
= 1.781 Mg/m^3

ρ_d = r/ (1 + w)
= 1.957/ (1+0.099)
= 1.781 Mg/m^3

Dry unit weight}

γ_d = g/ (1 + w)
= 19.20/ (1+0.099)
= 17.47 kN/m^3

FIELD MEASUREMENTS

Measurements taken in the field are mostly to determine density/unit weight. The most common application is the determination of the density of rolled and compacted fill, *e.g.* in road bases, embankments, etc.

Note: These methods are covered in detail by BS1377. You should understand the general principle that density is calculated from the mass and volume of a sample. How a sample of known volume is obtained depends on the nature of the soil. You are not expected to remember the details of each method

The Core Cutter Method

This method is suitable for soft fine grained soils.

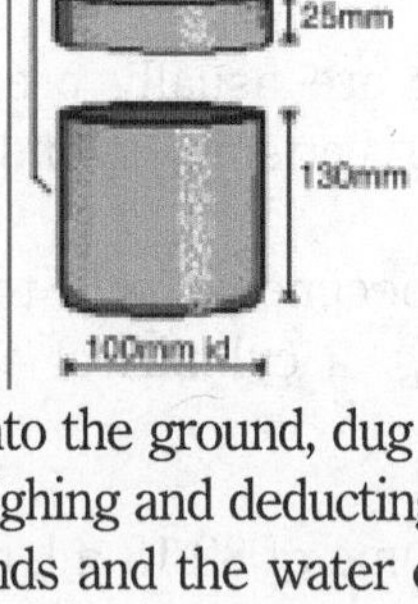

A steel cylinder is driven into the ground, dug out and the soil shaved off level. The mass of soil is found by weighing and deducting the mass of the cylinder. Small samples are taken from both ends and the water content determined.

The Sand-Pouring Cylinder Method

This method is suitable for stony soils

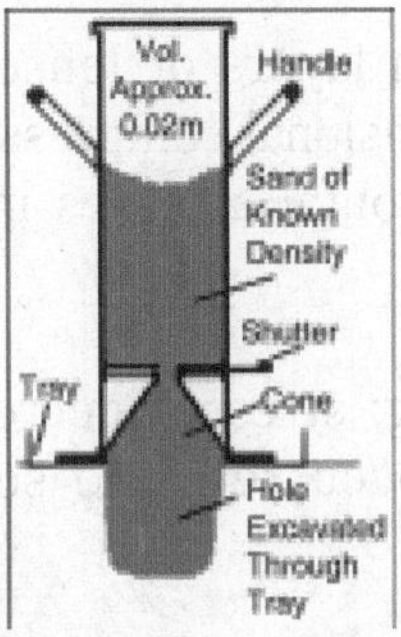

Using a special tray with a hole in the centre, a hole is formed in the soil and the mass of soil removed is weighed. The volume of the hole is calculated from the mass of clean dry running sand required to fill the hole. The sand-pouring cylinder is used to fill the hole in a controlled manner.

The mass of sand required to fill the hole is equal to the difference in the weight of the cylinder before and after filling the hole, less an allowance for the sand left in the cone above the hole as shown in figure 1.26.

Bulk Density

ρ = (Mass of soil)/(Volume of core cutter or hole)

CURRENT STATE OF SOIL

The state of soil is essentially the closeness of packing of the grains in the range:

Closely packed	→	Loosely packed
Dense	→	Loose
Low water content	→	High water content
Strong and stiff	→	Weak and soft

The important indicators of the current state of a soil are:

- *Current Stresses:* Vertical and horizontal effective stresses
- *Current Water Content:* Effecting strength and stiffness in fine soils
- *Liquidity Index:* Indicates state in fine soils
- *Density Index:* Indicates state of compaction in coarse soils
- *History of Loading and Unloading:* Degree of overconsolidation

Engineering operations (*e.g.* excavation, loading, unloading, compaction, etc.) on soil bring about changes in its state.

Its initial state is the result of processes of erosion and deposition. It is possible for the engineer to predict changes that could result from a proposed engineering operation: changes from the soil's *current state* to a new future state.

SOIL HISTORY: DEPOSITION AND EROSION

Original Deposition

Most soils are formed in layers or lenses by deposition from moving water, ice or wind. One-dimensional compression occurs as overlying layers are added. Vertical and horizontal stresses increase with deposition.

Erosion

Erosion causes unloading; stresses decrease; some vertical expansion occurs. Plastic strain has occurred; the soil remains compressed, *i.e.* overconsolidated.

Subsequent Changes

Subsequent changes may occur in the depositional environment: further loading/unloading due to glaciation, land movement, engineering; and ageing processes.

SOIL HISTORY: AGEING

The term *ageing* includes processes that occur with time, except loading and unloading. Ageing processes are independent of changes in loading.

Vibration and Compaction

Coarse soils can be made more dense by vibration or compaction at essentially constant effective stress

Creep

Fine soils creep and continue to compress and distort at constant effective stress after primary consolidation is complete.

Cementing and Bonding

Intergranular cementing and bonding occurs due to deposition of minerals from groundwater, *e.g.* calcium carbonate; disturbance due to excavation fractures the bonding and reduces strength.

Weathering

Physical and chemical changes take place in soils near the ground surface due to the influence of changes in rainfall and temperature.

Changes in Salinity

Changes in the salinity of groundwater are due to changes in relative sea and land levels, thus soil originally deposited in sea water may later have fresh water in its pores, such soils may be prone to sudden collapse.

DENSITY INDEX (RELATIVE DENSITY)

The void ratio of coarse soils (sands and gravels) varies with the state of packing between the loosest practical state in which it can exist and the densest. Some engineering properties are affected by this,

Shear Strength

Seepage

The flow of water through soil.

Seepage Force (J)

The force transmitted to a body of soil due to the seepage of groundwater. $J = i\,\gamma_w\,V$

Seepage Pressure (j)

The seepage force per unit volume $j = i\,\gamma_w$

Seepage Velocity (v_s)

The average velocity at which groundwater flows through the pores; the ratio of the volume flow rate to the average area of voids in a cross-section.

$$v_s = q/\,A_v$$

Sensitivity (S_t)

$$S_t = \frac{\text{Undisturbed undrained strength}}{\text{Remoulded undrained strength}}$$

A measure of the change in strength of f clays upon disturbance: For ordinary clays S_t = 1 to 4, sensitive clays 4 to 8, 'quick' clays 16–100.

Settlement (s, ρ)

The downward movement of ground or ground surface; the downward movement of a foundation.

Shape Factors ($s_c\ s_q\ s_g$)

Factors used in a general equation giving ultimate bearing capacity which provide adjustment relating to the shape (*e.g.* strip, square, circle)

Shear Modulus (G')

The ratio of the change in shear stress to the resulting change in shear strain.

Shear Strain (γ)

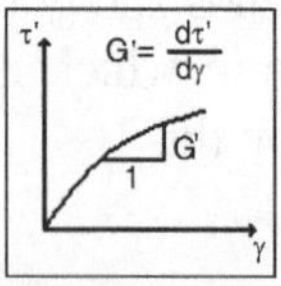

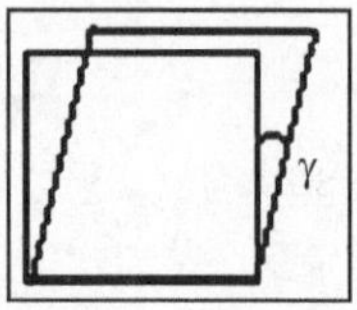

The angular distortion or change in shape of a body of material.

Shear Strength (τ_f)

The maximum shear stress which a material can sustain under a given set of conditions. In soil mechanics it is necessary to refer shear strength to the strain at which the strength is measured.

- *Critical shear strengt*: $\tau_c = c'_c + \sigma' \tan\phi'_c$
- *Peak shear strength*: $\tau_p = c'_p + \sigma' \tan\phi'_p$
- *Undrained shear strength*: $\tau_c = s_u$
- *Residual shear strength*: $\tau_r = c'_r + \sigma' \tan\phi'_r$

Shear Stress (τ)

The force per unit area acting tangentially to a given plane or surface. Units: kPa.

Short-term Conditions

Conditions in the ground when it is undrained; loading or unloading will cause changes in pore pressures, but not immediately in volume; these excess pore pressures dissipate with time due to consolidation.

Shrinkage Limit (w_s)

(Also SL) The water content below which further reduction in water content causes no further reduction in volume.

Skin Friction Stress (f_s)

The shear stress on the shaft of a pile of a pile or caisson or cone penetrometer. *Note:* Although known as 'skin friction', the shear stress may not vary with normal stress.

Soil Suction

Negative pore pressure created by capillary attraction in fine soils and in unsaturated soils.

Specific Gravity (G_s)

The ratio of the mass of a body or a substance to the mass of an equal volume of water; the ratio of the density of a body or a substance to that of water.

$$G_s = \frac{\text{Mass of body}}{\text{Mass of the same volume of water}}$$

Specific Surface (S_s)

The total surface area of all particles in a unit mass of soil. Units: m^2/g.

Specific Volume (v)

The total volume of a quantity of soil containing a unit volume of soil grains.

$$v = 1 + e$$

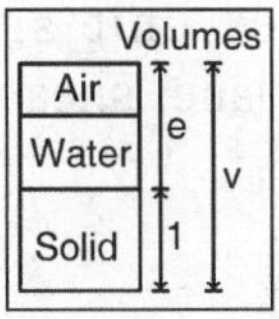

State Boundary Surface

The state boundary surface (SBS) is the boundary (usually in q':p':v space) to all possible stress-volume states of a soil.

If a soil with a state on the state boundary surface is unloaded, the subsequent state will be inside the SBS, upon re-loading the subsequent state will move back onto (but not beyond) the SBS. In other words, stable states cannot exist outside the SBS.

State Parameter (S_s, S_v)

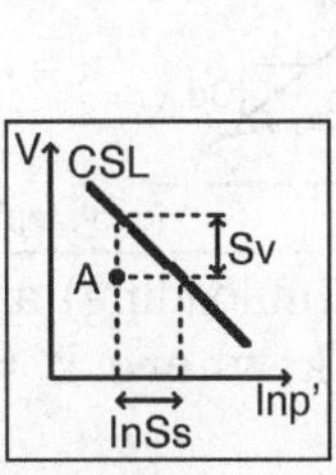

A measure of the distance between the current state (A) and the critical state; expressed as a ratio of stresses or as a difference of specific volumes,

$$S_s = p'_A / p'_c$$

$$S_v = v_A - v_c$$

Steady State Pore Pressure (u_o)

The pore pressure at equilibrium when all excess pore pressures have fully dissipated.

Stiffness

Susceptibility to distortion or volume change under load.

Strain

A measure of the change in size or shape of a body, relative to its original size or shape. (Direct strain is the ratio of change in length to original length; shear strain is the angle of distortion; volumetric strain is the ratio of change in volume to original volume.)

Stream Function (Ψ)

A function introduced in the solution of the Laplace equations defining two-dimensional seepage flow; the flow-quantity interval ($\Delta\theta$) between stream lines (flow lines) can be stated as $\Delta\Psi = \Delta\theta$.

Stress

The intensity of force per unit area; normal stress is applied perpendicularly to a surface or plane, shear stress is applied tangentially to a surface or plane.

Stress History

The past history of loading and unloading associated with a soil.

Summation (Σ)

The symbol Σ when placed in front of a quantity indicates that the quantity is a sum or total.

Swelling/Recompression Index (C_s, k)

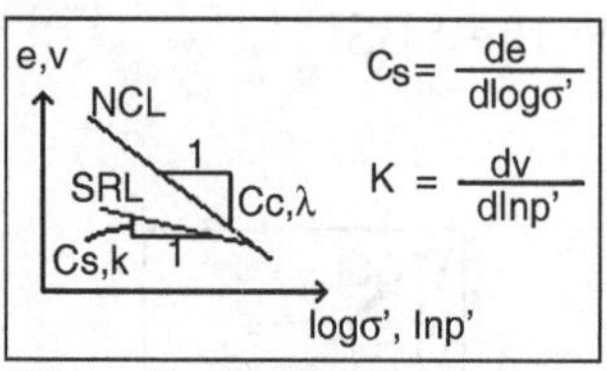

The slope of the swelling (unloading) and recompression (reloading)line. $C_s = 2.303$ k where k = slope of v:ln σ´ swelling/ recompression line.

Compressibility

Capillary Rise (h_c)

The height to which water will rise above the water table due to capillarity.

Clay

Soil particles less than 0.002 mm in size.

Coarse-grained Soils

Soils with less than 35% by mass of grains less than 0.06mm in size:

- *Coarse soil*: Grain size is predominantly 0.06–60 mm
- *Very coarse soil*: Grain size is predominantly greater than 60mm

Coefficient of Active Earth Pressure (K_a)

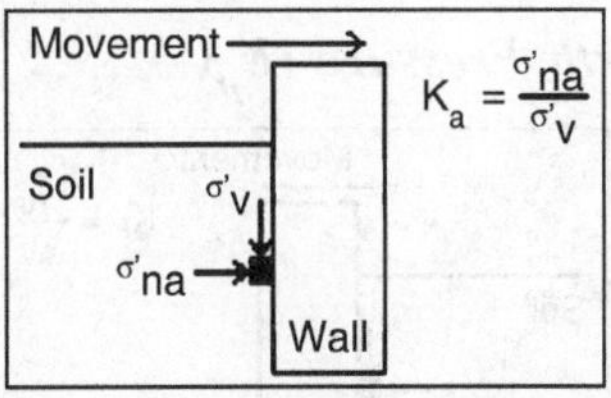

The ratio of the minimum horizontal effective stress to the vertical effective stress at a point in a soil mass retained by a surface as the surface moves away from the soil.

Coefficient of Consolidation (c_v and c_h)

$$C_v \frac{\partial^2 \bar{u}}{\partial z^2} = \frac{\partial \bar{u}}{\partial t}$$

A measure of the rate of change of volume during primary consolidation; referred to either vertical drainage (c_v) or horizontal drainage (c_h). Units: m^2/s

Coefficient of Curvature of Grading (C_z)

(Also curvature coefficient) A measure of the shape of a grading curve: $C_z = d_{30}{}^2/ (d_{60} \times d_{10})$.

Coefficient of Earth Pressure at Rest (K_o)

The ratio of horizontal effective stress to vertical effective stress at a point in a soil mass loaded in conditions of zero horizontal strain.

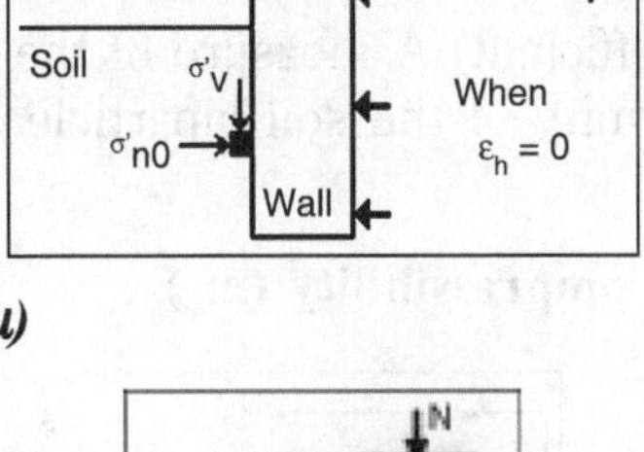

Coefficient of Friction (μ)

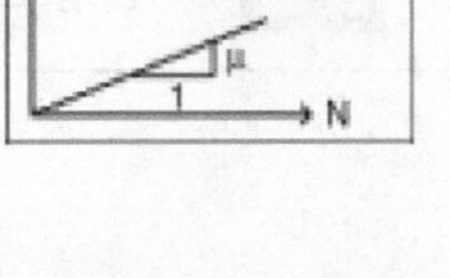

The ratio between the tangential force (T) required to cause a body to slide along a plane and the normal force (N) between the body and the plane: $T = \mu N$

Along a wall or foundation surface:

$$\mu = \tan \delta$$

where δ = angle of wall friction,

Coefficient of Passive Earth Pressure (K_p)

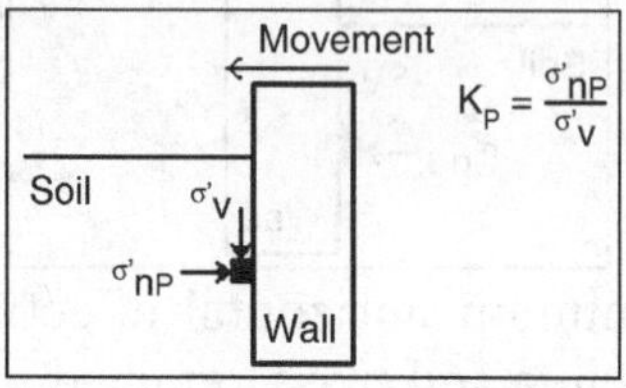

The ratio of the maximum horizontal effective stress to the vertical effective stress at a point in a soil mass retained by a surface as the surface moves towards from the soil.

Coefficient of Permeability (k)

The constant average discharge velocity (v) of water passing through soil when the hydraulic gradient (i) is 1.0; defined by Darcy's law: $v = k.i$

Coefficient of Secondary Consolidation (C_a)

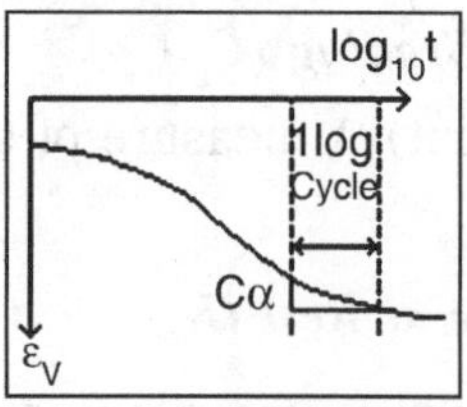

The change in volumetric strain per $\log_{10}$ cycle of time after primary consolidation is complete.

Coefficient of Uniformity of Grading (C_u)

(Also uniformity coefficient) A measure of the slope of a grading curve and therefore the uniformity of the soil inparticle size analyses: $C_u = d_{60}/d_{10}$

Coefficient of Volume Compressibility (m_v)

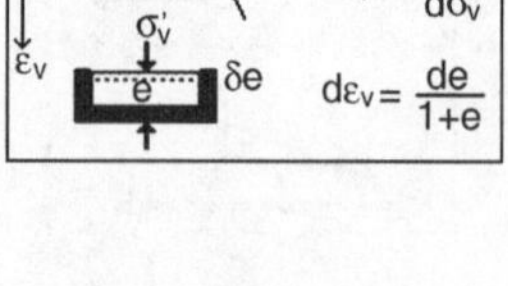

The change in volumetric strain per unit volume per unit change in effective stress in one-dimensional compression. (Units m²/MN) Compressibility is the reciprocal of stiffness.

Cohesion (c′)

Apparent cohesion (c′) is the intercept on the shear stress axis of a straight-line Mohr-Coulomb envelope. For critical states and residual states, $c'_c = c'_r = 0$ in most cases. For peak states, a curved strength envelope also passes through the origin, but a straight-line fitted to a small number of results is often extrapolated to give an intercept $c'_p > 0$. (In physics, cohesion is described as 'the force that holds together molecules or like particles within a substance'.)

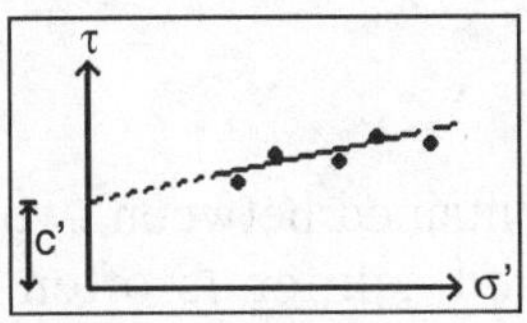

Compaction

Volume change in soil in which air is expelled, but with the water content remaining constant. Compaction may occur due to vibration in loose sands and gravels, and in fill due to self-weight. In soil constructions, compaction is achieved by rolling, tamping or vibrating.

Compatibility

The relationship between the strains in a deforming body so that no holes appear and no material is destroyed.

Compressibility of Pore Fluid (C_v)

$$C_v = \frac{d\varepsilon_v}{d\sigma}$$

The ratio of the change in volumetric strain to the change in isotropic stress. For a saturated soil, $C_v = \infty$

Compression Index (C_c)

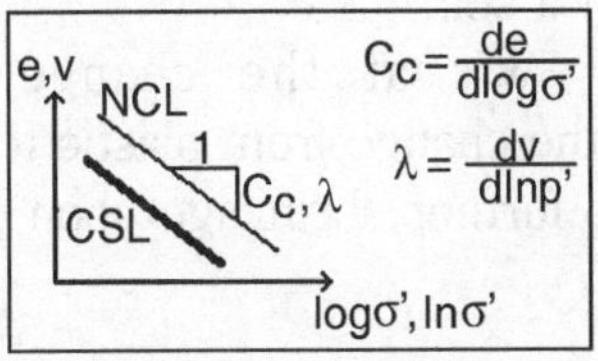

The slope of the normal compression line (NCL) and critical state line (CSL).

Cone Resistance (q_c)

The resistance force divided by the end area of the cone tip, measured during the cone penetration test.

Cone Penetration Test

A penetration test in which a sleeved cone (diameter 35.7mm, end area 1000mm^2, apex angle 60°) is pushed into the ground at a rate of 20 mm/min and the force required measured.

The magnitude of this force divided by the end area is called the *cone penetration resistance* (q_c).

The cone and the sleeve can be advanced separately and so the frictional resistance along the sleeve (1000mm^2 area) or *local side friction* (f_s) can also be measured.

Confined Aquifer

A *confined* aquifer is contained between two strata of low permeability; in flow analyses, a confined aquifer is often assumed to be saturated throughout its depth.

Consistency Index (I_c)

A measure of the relationship between the current water content and the consistency limits, similar in form to the density index, but expressed in terms of water content.

$$I_c = (w_L - w)/I_P$$

- At liquid limit, $I_c = 0$
- At plastic limit, $I_c = 1.0$

Consistency Limits

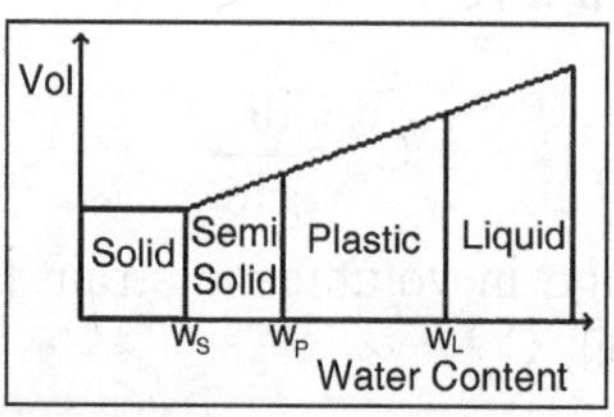

(Also called Atterberg limits) Measures of water content corresponding to changes in physical state of a soil:

liquid limit (w_L) = w/c at the change from liquid to plastic plastic limit (w_P) = w/c at the change from plastic to semi-solid shrinkage limit (w_s) = w/c below which no further shrinkage upon drying occurs

Consolidation

Volume change due to dissipation of excess pore pressure, usually with constant total stress.

Consolidation Settlement (s_c, ρ_c)

The settlement of a foundation due to consolidation.

Coulomb's Equation

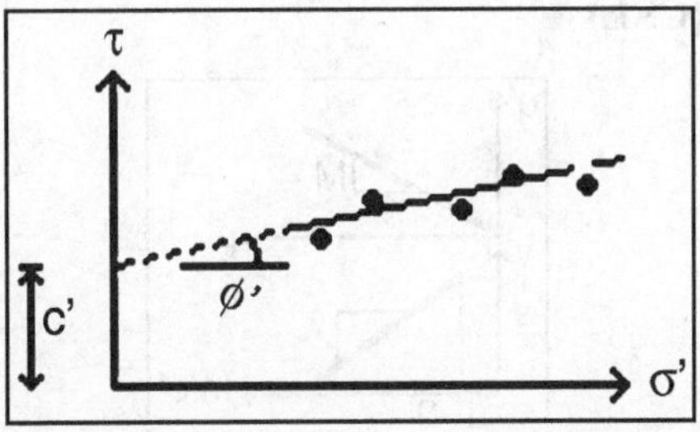

(After Charles Augustin Coulomb, 1736–1806) An equation relating the shear strength of soil to the normal effective stress on a failure plane.

$$\tau_f = c' + \sigma' \tan\phi' = c' + (\sigma - u) \tan\phi'$$

Creep

Deformation or volume change which occurs in soil at constant effective stress progressing with time.

Critical Circle

In slope stability analyses the slip circle corresponding to the lowest factor of safety.

Critical Ground Slope Angle (i_c)

The ground slope angle that corresponds to a slope-stability factor of safety of 1.0.

Critical Height (H_c)

The height of a slope (*e.g.* embankment, cutting, trench) for which the factor of safety against collapse is 1.0.

Critical Hydraulic Gradient (i_c)

The hydraulic gradient at which effective stresses becomes zero; with upward seepage, sand may become quicksand.

Critical Shear Strength (τ_c)

The shear stress developed along a slip surface during shearing at constant volume; also known as ultimate shear strength.

Critical State

The state of a soil in which it strains; critical states occur on the critical state line.

Note: The critical state is not the same as the residual state; at the critical state, soil particles continue to rotate and the flow is turbulent; at the residual state, in clay soils, flat particles become aligned to the slip plane and the flow is laminar.

Critical State Line (CSL)

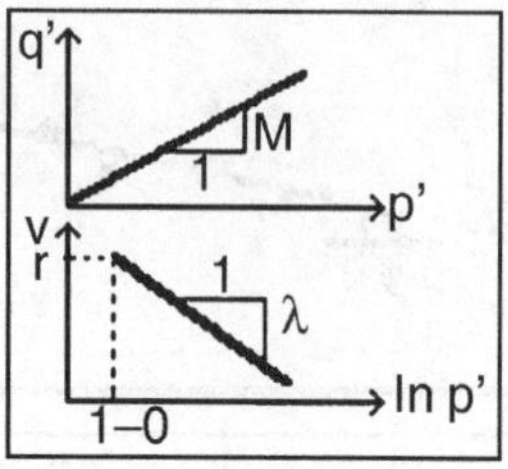

The unique relationship at failure between deviator stress, average normal stress and volume (or shear stress, normal stress and void ratio) is defined by the critical state line.

$$q' = Mp'$$
$$v = \Gamma - \lambda \ln p'$$
$$\tau' = \sigma'_n \tan\phi'$$
$$e = e_G - C_c \log\sigma'_n$$

Critical Void Ratio (e_c)

The void ratio of a soil at which its volume remains constant during shearing. Note: critical void ratio depends on the mean stress.

Current State of Soil

The current state of a soil is described by its voids ratio (e) or specific volume (v), the current stress (σ' or p') and the overconsolidation ratio or yield stress ratio (R_y).

Ppermeability

The property which allows the flow of water through a soil.

pF Index

A measure of soil suction: pF = $\log_{10}$ (suction head in cm),*i.e.* for a suction of 100 kPa, pF = 2.0 [Range for soils is pF = 0 to 7]

pH Value

A measure of acidity or alkalinity of groundwater or soil water extract based on the hydrogen ion content:

- pH = $-\log_{10}$(hydrogen ion content)

- pH < 7.0 indicates acidity.
- pH > 7.0 indicates alkalinity.

Piezometer

An instrument used to measure *in situ* pore pressures; may be an open standpipe or an enclosed electronic pressure transducer.

Piezometric Surface

An imaginary surface corresponding to the hydrostatic water level of a confined body of groundwater; the notional level to which artesian pressure would raise water in a (real or imaginary) standpipe.

Pile Spacing (s)

The distance from centre to centre of piles in a group.

Plane Strain

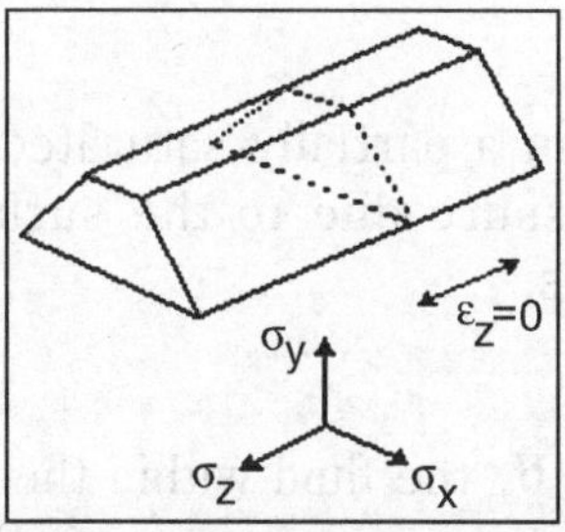

A two-dimensional stress state, where the out-of-plane strain (*i.e.* the strain normal to the plane being considered, e_z) is zero. An example of a plane strain situation would be on a cross-section through a long structure being loaded in the x-y plane, such as an embankment dam.

Plastic Deformation

The flow or distortion resulting in a permanent and irrecoverable change in shape or volume.

Plastic Limit (w_p)

(Also PL) The moisture content above which a soil will have a plastic consistency, but below which it crumbles.

Plastic Strain

Deformation or strain that is not recovered upon unloading.

Plasticity

1. The property of a soil (or other material) which allows it to deform continuously.

2. Plasticity theory is used to calculate plastic (irreversible) deformations.

Plasticity Index (I_p)

(Also PI) The difference between the liquid limit and plastic limit. $I_p = w_L - w_P$

Poisson's Ratio (ν, ν')

The ratio of the change in strain perpendicular to the direction of loading to the change in strain caused in the same direction.

$\nu' = -d\varepsilon_y / d\varepsilon_x$ *(for to loading or unloading in the x-direction).*

For undrained loading of saturated soil,

$$\nu_u = 0.5$$

For drained loading or unsaturated soil,

$$\nu' = 0.2\text{–}0.5$$

Pore Air Pressure (u_a)

The pressure of air in a partially saturated soil; not necessarily the same as pore water pressure due to the surface tension on air-water interfaces within the voids.

Pore Pressure (u)

The pressure exerted by the fluid within the pores or voids in a porous material; in saturated soil the pore pressure is the pore water pressure.

Pore Pressure Coefficient (A)

The ratio of the change in pore pressure to the change in deviator stress, *e.g.* in an undrained triaxial test; the value of A varies with strain and the overconsolidation ratio.

Pore Pressure Coefficient (B)

The ratio of the change in pore pressure to the change in isotropic stress in undrained loading.

$$B = \frac{\Delta u}{\Delta p}$$

For saturated soils, B = 1.

Pore Pressure Force (U)

The resultant force due to pore pressure acting on a given area.

Pore Pressure Ratio (r_u)

$$r_u = \frac{u}{\sigma_v}$$

At a given point in a body of soil, the ratio of the porewater pressure to the vertical overburden pressure.

Pore Water Pressure (u_w)

Porosity (n)

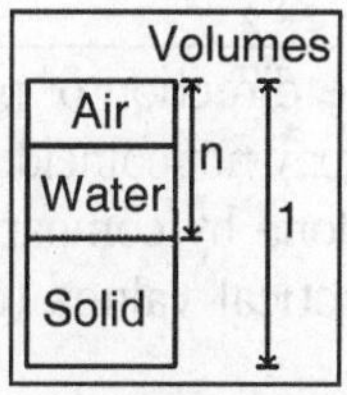

The ratio of void volume to total volume:

$$n = V_v / V$$

where,

$$V_v = \text{Volume of voids}$$
$$V = \text{Total volume}$$

Potential Function (Φ)

A function introduced in the solution of the Laplace equations defining two-dimensional seepage flow.

Pressure Head (h_w)

(Also head) The height of a column of water required to develop a given pressure u at a given point.

$$h_w = u / \gamma_w$$

Pressure in Tension Crack (p_w)

The horizontal pressure exerted in a slope or against a retaining wall due to hydrostatic water pressure in tension cracks.

Principal Axes

A set of orthogonal axes perpendicular to which the shear stresses and shear strains are zero and normal stresses and strains are referred to as principal stresses and principal strains.

Principal Strains (ε_1, ε_2, ε_3)

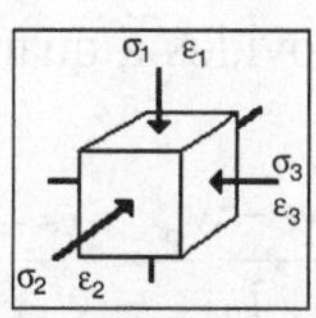

The strains occurring in the directions of the principal axes of strain. Note: the principal axes of stress and strain may not coincide.

Principal Stresses (σ_1, σ_2, σ_3)

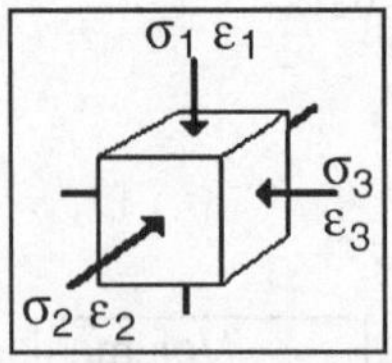

Normal stresses acting in the direction of principal axes of stress. Note: the principal axes of stress and strain may not coincide. It is therefore useful to measure the *in situ* state and this can be done by comparing the *in situ* void ratio (e) with the minimum and maximum practical values (e_{min} and e_{max}) to give a density index I_D

$$I_D = \frac{e_{max} - e}{e_{max} - e_{min}}$$

e_{min} is determined with soil compacted densely in a metal mould e_{max} is determined with soil poured loosely into a metal mould

Table. Density Index is Also known as Relative Density Relative States of Compaction are Defined:

Density index	State of compaction
0-15%	Very loose
15-35	Loose
35-65	Medium
65-85	Dense
85-100%	Very dense

LIQUIDITY INDEX

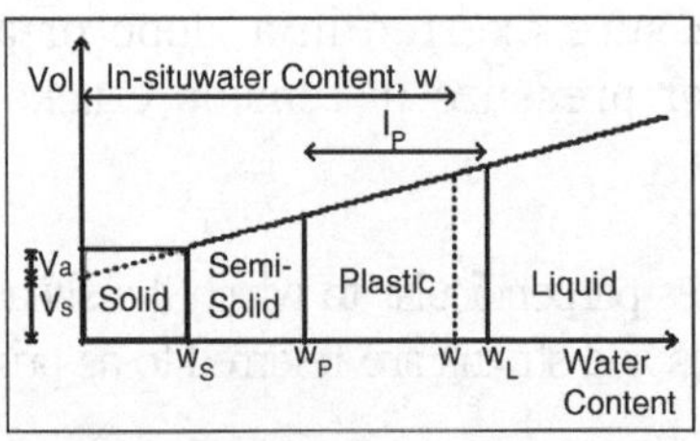

In fine soils, especially clays, the current state is dependent on the water content with respect to the consistency limits (or Atterberg limits). The liquidity index (I_L or LI) provides a quantitative measure of the current state:

$$I_L = \frac{w - w_P}{I_P} = \frac{w - w_P}{w_L - w_P}$$

where,

$$w_P = \text{plastic limit and}$$

$$w_L = \text{liquid limit}$$

Significant values of I_L indicating the consistency of the soil are:

$$I_L < 0 \Rightarrow \text{semi-plastic solid or solid}$$

$$0 < I_L < 1 \Rightarrow \text{plastic}$$

$$1 < I_L \Rightarrow \text{liquid}$$

PREDICTING STIFFNESS AND STRENGTH FROM INDEX PROPERTIES

Preliminary estimates of strength and stiffness can provide a useful basis for early design and feasibility studies, and also the planning of more detailed testing programmes. The following suggestions have been made; they are simple, but not necessarily reliable, and should be not be used in final design calculations.

Undrained Shear Strength

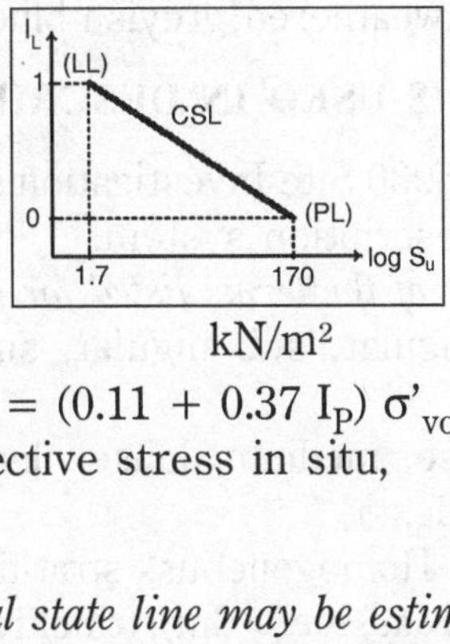

kN/m²

$$s_u = (0.11 + 0.37\, I_P)\, \sigma'_{vo} \ \text{kN/m}^2$$

where σ'_{vo} = vertical effective stress in situ,

Stiffness

The slope of the critical state line may be estimated from:

$$\lambda = I_P.G_s/ 461$$

The compressibility index may be estimated from:

$$C_c = \lambda \ \text{In}10 = I_P\, G_s/ 200$$

(where P is in percentage units)

THE COMPRESSIBILITY INDEX MAY BECLASSIFICATION

BS 5930 Site Investigation recommends the terminology and a system for describing and classifying soils for engineering purposes. Without the use of a satisfactory system of description and classification, the description of materials found on a site would be meaningless or even misleading, and it would be difficult to apply experience to future projects.

BS DESCRIPTION SYSTEM

A recommended protocol for describing a soil deposit uses

ninecharacteristics; these should be written in the following order:

- *Compactness*: *e.g.* loos , dense, slightly cemented
- *Bedding structure*: *e.g.* homogeneous or stratified; dip, orientation
- *Discontinuities*: spacing of beds, joints, fissures
- *Weathered state*: degree of weathering
- *Colour*: main body colour, mottling
- *Grading or consistency*: *e.g.* well-graded, poorly-graded; soft, firm, hard
- *Soil Name*: *e.g.* Gravel, Sand, Silt, Clay; plus silty, gravelly-, with-fines, etc. as appropriate
- *Soil class*: (BSCS) designation (for roads and airfields) *e.g.* SW = well-graded sand
- *Geological stratigraphic name*: (when known) *e.g.* London clay

Not all characteristics are necessarily applicable in every case.

Example:

- Loose homogeneous reddish-yellow poorly-graded medium SAND (SP), Flood plain alluvium
- Dense fissured unweathered greyish-blue firm CLAY. Oxford clay.

DEFINITIONS OF TERMS USED IN DESCRIPTION

A table is given in BS 5930 Site Investigation setting out a recommended field indentification and description system.

The following are some of the terms listed for use in soil descriptions:

- *Particle shape*: Angular, sub-angular, sub-rounded, rounded, flat, elongate
- *Compactness*: Loose, medium dense, dense (use a pick or driven peg, or density index)
- *Bedding structure*: Homogeneous, stratified, inter-stratified
- *Bedding spacing*: Massive(>2m), thickly bedded (2000-600 mm), medium bedded (600-200 mm), thinly bedded (200-60 mm), very thinly bedded (60-20 mm), laminated (20-6 mm), thinly laminated (<6 mm).
- Discontinuities: *i.e.* spacing of joints and fissure: very widely spaced(>2m), widely spaced (2000-600 mm), medium spaced (600-200 mm), closely spaced (200-60 mm), very closely spaced (60-20 mm), extremely closely spaced (<20 mm).
- *Colours*: Red, pink, yellow, brown, olive, green, blue, white, grey, black
- *Consistency*: Very soft (exudes between fingers), soft (easily mouldable), firm (strong finger pressure required), stiff (can be indented with fingers, but not moulded) very stiff (indented by sharp object), hard (difficult to indent).

- *Grading*: Well graded (wide size range), uniform (very narrow size range), poorly graded (narrow or uneven size range).
- *Composite soils*: In Sands and Gravels: slightly clayey or silty (<5%), clayey or silty (5-15%), very clayey or silty(>15%) In Clays And Silts: sandy or gravelly (35-65%)

2

Maintaining Soil Fertility Under an Organic Management

Soil, one of the most important natural resources, is the medium in which our food, clothing and shelter are produced. Soil is a layer of organic and mineral substances covering most of the Earth's surfaces. It is created by a slow, and constant physical and chemical breakdown of rock and the action and turnover of living organisms. It takes approximately 500 years to replace 1 inch of soil by these processes, making it critical to keep the soil fertile and productive as well as to prevent its erosion. Soil has four main components: mineral, air, water, and organic matter. Minerals comprise 45% of the soil, water and air 50%, and organic matter 2% to 5%. The organic component consists of dead organisms, plant matter and other organic materials in various stages of decomposition.

The organic component is important for nutrition because it serves as a reservoir of nutrients for plants and provides nutrients for soilborne organisms. When the organic matter is stable and no longer undergoing decomposition it is called humus. Microscopic organisms are essential to the stabilization of humus and the release of nutrients from organic matter. The release of nutrients from the soil to the plant is an interdependent process relying on the organisms that reside in the soil. These soil-dwelling organisms include earthworms, arthropods, bacteria, fungi, algae, and nematodes. A soil high in organic matter and humus is characterized as having good tilth or soil structure. This type of soil is easy to work and is in overall good health, which encourages the processes necessary to produce a healthy plant.

SOIL MANAGEMENT

The foundation of organic farming lies in the health of the soil. A fertile soil provides essential nutrients to a growing crop plant and helps support a diverse and active biotic community. Strategies the transitional farmer will employ to build the soil are crop rotations, animal and green manures, and cover cropping.

ANIMAL MANURES

Livestock manure traditionally has been used to fertilize soils of both organic and sustainable farms. Manure can be applied to the field in either raw or composted form. The organic farmer must follow the Final Rule's specific requirements for using raw or composted manure. Raw manure is useful in supplying nutrients and adding organic matter to the soil, as well as encouraging healthy biological processes in the soil. However, it is important to know what is in the manure because some may contain contaminants, and nutrient content varies with the animal, bedding, and storage. It is a good idea to compost manure, because the heat created during composting may kill weed seeds and break down contaminants such as antibiotics. Fresh manure generally will have higher available nitrogen, but overapplications can lead to salt buildup and leaching. A soil test to monitor soil fertility is highly recommended so the farmer can add the right amount of raw or composted manure to plots and avoid nutrient imbalances.

COVER CROPPING

A cover crop can be an annual, biennial, or perennial herbaceous plant grown in a pure or mixed stand for all or part of the year. This crop provides soil cover and can help loosen compacted soil through the growth of roots and improved water filtration.

Cover crops can maintain or increase soil organic matter if they are allowed to grow long enough to produce high herbage. Cover crops also help prevent soil erosion caused by both water and wind, suppress weeds by keeping the sun from reaching weed seeds, and reduce insect pests and diseases. In addition, a legume used as a cover crop can provide nitrogen to the soil. Non-legumes can take up excess nitrogen, phosphorus, and potassium from previous crops and recycle them to the following crop. Common cover crops and green manures include rye, buckwheat, hairy vetch, crimson clover, subterranean clover, red clover, sweet clover, cowpeas, millet, and forage sorghums.

GREEN MANURES

When a cover crop is tilled into the soil while it is still lush and green, it is referred to as a green manure. Green manures are important under an organic farming system because they help to add organic matter and nutrition to the soil. When a green plant is incorporated into the soil, it has high amounts of nitrogen and moisture and becomes a food source for soil microorganisms and earthworms.

During the process of decomposition by the organisms in the soil, organic matter and nutrients become available to the crop plants. An additional benefit from using green manures is the suppression of weeds and soilborne diseases.

CROP ROTATIONS

Many farmers find that rotating crops improves the tilth or aggregation of the soil. Planning a crop rotation requires a farmer to plant crops at different times and in different locations in the same field. Usually, the succeeding crop will be of a different variety and species than the previous crops. Crop rotations can also be used to promote the soil's fertility, reduce erosion, reduce the buildup of pests, and spread out financial risk in case a crop fails. Farmers who include a legume in the rotation can increase the availability of nitrogen in the soil. The rhizobia that form on the nodules of legume roots convert nitrogen from the atmosphere into organic nitrogen, which that then becomes available to plants.

REQUIRED ELEMENTS FOR PLANT NUTRITION

Sixteen elements are considered necessary for plant growth. Carbon, hydrogen, and oxygen make up most of the elements in plants and are used in the formation of organic compounds. The other 13 elements are classified as macroor micronutrients based on the amount found in plants. Macronutrients of greatest concern are nitrogen, phosphorus, and potassium. Other macronutrients include sulfur, calcium, and magnesium. Micronutrients include iron, zinc, manganese, copper, boron, chlorine, and molybdenum. In general, fertilizers or amendments for organic production must come from natural carbon containing Non-synthetic materials or Non-synthetic inorganic materials. Organic fertilizers usually are needed in larger quantities than conventional fertilizers since they are lower in nutrients. It is good to know the nutrient analysis of the organic fertilizer before its application so that the right amount is applied.

- *Nitrogen:* The most common sources of nitrogen are from animal manure, either composted or raw, and green manures in the form of legumes that fix nitrogen from the atmosphere. Additional sources of nitrogen include blood meal, fish emulsion, fish protein, kelp and seaweed, and various vegetable meals.
- *Phosphorus:* Sources include manure, bone meal, fish and poultry meal, and rock phosphate.
- Potassium: Sources include manure, alfalfa meal, kelp meal, greensand, wood ash, potassium sulfate, and granite dust.
- *Sulfur:* This element is found in rainwater, especially in acid rain. When a farmer uses organic sources of nitrogen, the sulfur requirements are usually met.
- *Calcium:* Generally supplied through lime applications used to adjust acidic soils. Sources include calcitic lime, and dolomitic lime that also contains magnesium. Other sources include colloidal phosphate, bone meal, gypsum, and wood ashes.

- *Soil pH :* It is important to manage the pH of the soil since it can affect the plant's ability to take up nutrients and the microbial activity in the soil that affects the processes needed for plant nutrition. To correct acidic soil, farmers apply lime based on a soil test. An alkaline soil can be more difficult to correct since adding sulfur to make the soil more acidic is only a temporary remedy and can be more expensive than liming. When adjusting the pH, it is important to know the crop's pH requirement since different crops grow best at different pH levels. For most crops, optimum pH levels are between 6.0 and 7.0.

SOIL TEST

Using soil tests and monitoring crop growth will indicate to the farmer what nutrients need to be added to the soil. The soil test is often required for certification. Tissue analysis can complement the soil analysis and offer more clues about the fertilizers that need to be added. There are three sources available to a farmer or grower to get a soil and tissue analysis: a home testing kit, a government or university lab, and a private laboratory.

SOIL EROSION

To keep soil from eroding, farmers employ contour plowing and planting around hillsides instead of planting up and down hills. Strip cropping, in which a row of clover or a row of grasses is planted in strips between crops, can be used to hold water and keep the soil from eroding. Farmers in hilly or mountainous terrain concerned with erosion can terrace their land. Terracing involves building small level patches into the slope and building a wall to keep the soil from being washed away. On treeless plains, farmers plant shelterbelts or wind belts to break the wind speed and decrease wind erosion. Managing crop residues and preparing seedbeds for planting through tillage can damage the soil structure and lead to an increase in erosion and the loss of organic matter. Tillage smooths the soil surface while incorporating plant residue. In the process, it destroys the soil structure as well as earthworm channels. When the soil structure is damaged, there is a decrease in pore space, which means a decrease in air and water infiltration. Many farmers are switching to conservation tillage methods to manage crop residue and prepare seedbeds by preserving the residue from the previous crop and reducing the number of times equipment passes over a field.

SOIL PRODUCTIVITY MANAGEMENT

- *Building Healthy Soils:* The important principle is 'healthy soils yield healthy crops'. The health of the soil has much to do with soil organisms [flora and fauna] that live on organic matter. This vital component takes care of modifying soils physically, chemically and biologically for growing healthier crops. Therefore sustained productivity of soils is the result of biological and physical properties of the soil but not vice versa.

- *Soil Organisms:* The crop growth, resistance to pests attack and yields are dependent on the diversity and populations of soil organisms. Upon the usage of chemicals in the soil, the soil organisms have been lost or their activities are suppressed. Now reviving the soil life base is crucial.
- *Biofertilisers:* It is important to know that augmenting few soil beneficial microorganisms in sick soil conditions like compacted soil, bad pH, presence of chemical residues, harmful organisms and little organic matter etc. is of little use. The soil organisms can flourish only where there is organic matter primarily along with moisture and air. The introduced microbes may compete with their own native species trying to suppress each other. It is better to provide congenial conditions for the native soil organisms to flourish.
- *Fight Diseases:* Soil organisms not only supply the plant nutrients but also protect the plants from harmful organisms. The kind of practices received by the soil reflects the kind of micro-organisms inhabiting it. It is observed that those soils that receive chemicals as nutrients tend to decrease the microbial diversity and only the hardy harmful microbes that are harmful to plants tend to dominate the soils. Whereas the practices like diverse cropping, organic ammendments etc increase the diversity of soil organisms and also improve their numbers.

MAINTAIN SOIL PRODUCTIVITY

On the balance of experimental evidence, those cropping elements that help to maintain soil productivity in parts of Africa, Asia and Australia have been emphasised.

These include:

- The need to retain organic matter and soil structure.
- The need for appropriate tillage which, in many situations, involves minimal soil disturbance.
- The need to replenish the nutrients removed in grain, straw and livestock products.
- The essential need to maximize ground cover.
- The desirability of mixing crop species, particularly to include legumes with effective nodulation (grain crops, forage crops, trees and shrubs).

- The need for any fallow period to be managed rather than to be left bare to be exploited by livestock, or for weed seeds and pests to accumulate in the soil.

The question arises: how are these conceptual elements best exploited in crop rotations? A framework, and the dynamic, and recurring, nature of decisions involving the choice of rotations and techniques for the maintenance of soil productivity.

The value of effectively-nodulated legumes (in this case, a pasture ley) in increasing soil nitrogen, as long as the nitrogen is not removed as a product. The associated increases in soil rhizobia and vesicular arbuscular miccorhizae under undisturbed pasture legumes. The impact of plant architecture and ground cover on soil loss and mineral leaching.

All the above give estimates of the benefits, to soil biology and chemistry, of the crop elements in a rotation. The aspects they illustrate are reflected by some Australian farmers' practices, under which the rotations chosen appear to maintain soil productivity though soil nitrogen levels at equilibrium may be only half those under continuous legume pasture. However, there are very few long-term studies which quantify the effects of different crops on soil conditions. Indeed, such studies seem to be a priority to encourage changes in cropping systems. For example, there are few data on nitrogen fixation by legumes on farms in Africa and Asia, and no known studies, that compare the effects of groundnuts on soil structure and fertility (which are possibly degrading) with other legumes such as soybeans.

There is little information on the effects of fallow management on soil biology, particularly on weed seeds, pests and diseases. On many farms, weeds spread uncontrolled during the fallow period, while experiments often unrealistically contrast weed-free extremes. Caporali and Onnis, in re-affirming the importance of crop/legume rotations, interestingly also show that weed density increased more during ten years of sunflower-based cropping under sunflower monoculture than under a sunflower-lucerne rotation. They show too that there are more weeds after mechanical- than chemical-control.

There is much information available on the effects of tillage and litter/stubble treatments on soil properties. The data, however, are not always consistent. This is not surprising as the timeliness of an operation is relevant to its effects on the soil and there are rapid changes in soil properties in the days following sowing. Moreover, several years of treatment are often needed before there are lasting effects upon soil characteristics that influence crop growth, for example, root length density.

Notwithstanding the variation, and at times the ephemeral effects of management on soil properties, it is clear that there are benefits from those crop rotations that enable crop residue retention. Three examples follow.

Gill and Aulakh show that zero tillage for three years gave the lowest soil bulk density and highest crop yields. Radford *et al.* found that zero tillage plus stubble removal gave the least soil water storage but that zero tillage plus stubble retention gave the greatest soil water capacity.

Chan *et al.*record that ten years, conventional (three-pass) cultivation and stubble burning gave least soil organic matter whilst direct drilling plus stubble retention gave most organic matter. In this study, tillage had a cumulative effect, causing 31 per cent differences in organic matter (2.42 per cent as against 1.68 per cent organic carbon); a loss of 1 per cent organic carbon was calculated to result in a loss of almost 3 cmole of negative charge per kilogram of soil. As a working hypothesis, it seems that stubble and litter retention (and incorporation if necessary to avoid removal by grasing or wind) is the key to sustainable rotations. Bearing in mind soil organisms, however, it may be necessary occasionally to till and mix the soil layers.

Mineral nutrition is, together with crop choice, stubble/litter treatment, and soil management, a key to sustainable rotations. For example, the removal of a 600 kg grain crop requires 1.8 kg phosphorus to be replaced using manure or mineral fertilizer. Nine kilograms of nitrogen are also needed (most realistically, from the breakdown of a previous legume crop or pasture, or by application of manure). A long fallow, in which deep-rooted perennials draw nutrients from depth, may help to balance the loss of nutrients. However, nitrate may move down the profile during the fallow, commonly giving a net annual decline in soil nitrogen in the absence of legumes, however, found that there was an increase (of 30 kg N/ha) during a short (7-8 months) fallow which balanced the loss during the crop phase. If weeds or pasture grow during the fallow and are grazed, then nutrient losses of about 20 per cent are common. These losses are by removal of animal products and volatilisation of urine. Grasing can also redistribute nutrients as dung from one field to another.

SOIL PROPERTIES AFFECT PLANT NUTRITION

Soil is a complex mixture of non-living substances (minerals, organic matter, gases and liquids) and living organisms (bacteria, fungi, insects, worms, etc.). These factors influence soil fertility either directly or indirectly.

Soil solids consist of mineral particles, organic matter in varying stages of decomposition and living organisms. Solids make up about half the soil volume, while water and gases make up the other half in the pore space.

Soil mineral particles vary widely in size and are classified by size:

- Rocks are larger than 2 mm (0.08") in diameter
- Sand particles range from 0.05 to 2 mm (0.002 to 0.08") in diameter
- Silt particles range from 0.002 to 0.05 mm (0.00008 to 0.002") in diameter

- Clay particles are smaller than 0.002 mm (0.00008") in diameter

These particles are made from various mineral types with different elemental composition, which affects weathering processes and thus the release of certain nutrients. Two soils with identical texture could be drastically different in fertility due to differences in mineral composition. Potassium is an example of a plant nutrient whose supply arises from mineral weathering in soil.

The soil colloidal fraction refers to microscopic particles of clay and organic matter. The surface of the colloidal fraction is where most soil chemical reactions occur and it is very important in nutrient supply.

The proportion of sand, silt and clay determines the texture of a soil. Soil texture is grouped into five or more classes. The texture influences fertility by affecting moisture holding capacity, air exchange and the CEC. Adequate moisture is key to fertilizer response and potential yield for canola in western Canada.

The CEC is an important property that influences the soil storage of many plant nutrients. Most nutrients are present in the soil water as positively charged cations. A few are negatively charged anions. The CEC indicates a soil's ability to hold or store cations. Prairie soil particles typically have a negative charge. The process of electrical attraction that holds cations to negative surfaces of soil colloids is called adsorption (not absorption). The cations are not permanently stuck to the colloidal surface and can be exchanged with other cations. With time certain cations may become 'fixed' • into forms that are not easily removed from the exchange complexes. Adsorbed cations are not removed by water moving through the soil and can be accessed by plant roots. Cations with a higher positive charge (for example Ca^{+2}) are held more tightly than those with a lower charge (for example K^{+}).

Soil negative charges arise due to substitutions in the mineral crystals by elements with smaller positive charge, and due to reactions at the edges. Organic particles also contain a significant number of negative charges. The total particle surface area in a soil increases as particle sizes get smaller. Therefore, a soil high in clay has a much greater surface area than a sandy soil.

A high clay soil also has a bearing on the surface area and negative charge. Clay minerals are microscopic layers of aluminium and silicon crystals formed by weathering of other minerals. Thus clays are called secondary minerals. The type of clay depends on the original minerals and the weathering extent. Fairly young clays common in western Canada (such as montmorillonite) have a 2:1 arrangement of silica:alumina crystal sheets, while older clays have a 1:1 arrangement. Generally, 2:1 clays have 10 to 100 times more surface area, negative charges and, consequently, a higher

CEC than 1:1 clays. Organic colloids have 10 to 100 times more negative charges and higher CECs than the 1:1 clays. Therefore, soil organic matter levels greatly influence the CEC.

The CEC strongly influences soil fertility. A higher CEC means that more cations, including plant nutrients, can be loosely stored in a plant available form, giving the plant a greater pool of nutrients to draw from. Since most cations are not highly soluble, only small quantities can be dissolved in the soil solution at one time. The CEC soil property allows a reservoir of nutrients to be stored then released to plant roots. This continuous replenishment of nutrients in soil water is very important for several nutrients, including potassium. A high CEC also means that fewer cations will be lost through leaching out of the root zone.

Since soils are predominantly negatively charged, anions [such as nitrate (NO_{3-}) and sulphate (SO_4^{-2})] are repelled by soil colloids and tend to stay in the soil water. They will flow with water and are potentially subject to leaching loss.

Soil organic matter (OM) plays an important role in soil fertility as a plant nutrient storehouse. Not only does OM adsorb many cations due to a high CEC, it also stores nutrients as part of its structure. As the OM is decomposed by soil microbes, nutrients are released from the organic structure into plant available forms this process is called mineralization. Mineralization from OM is the primary natural source of plant available N and S in prairie soils, and also influences P availability. Mineralization of individual nutrients will be described in later sections. Soil OM also plays a secondary role in soil fertility by improving physical properties such as water holding capacity, infiltration, aggregation (tilth) and buffering pH.

SOIL TESTING FOR NUTRIENT CONTENT

Plant nutrient content in soil varies over years, between fields and even within fields that appear uniform. Soil sampling and analysis methods (soil testing) were developed to assess the fertility level and to predict crop response to applied fertilizer or manure. Soil testing is not an exact science due to nutrient variability inherent in most fields and the inability to predict growing season weather. Although soil testing is not exact, it can help estimate soil fertility and give reasonable guides for profitable fertilizer application.

Spatial nutrient variability in fields creates problems for soil testing and fertilizer application. The variability makes it difficult to obtain representative soil samples. Using single fertilizer rates across variable fields results in over-fertilized and under-fertilized areas within the field. Although variable rate fertilization is being researched and developed, most fields still are fertilized with a single rate. In addition, fertilizer response calibrations developed from research sites with low variability will under-predict the optimum fertilizer rate for larger farm fields with more variability.

For meaningful soil test results proper soil sampling is necessary. A sampling error in the field is usually much greater than the analytical error in the lab. Ensure soil samples accurately reflect the overall field. However, intensive soil sampling is not convenient, cost effective or practical. Research shows that the accuracy of a composite soil sample increases with the number of sub-samples taken. Accuracy refers to how similar the soil sample value is to the true field average (is the sample representative). Precision describes how often the same value can be obtained when repeating the procedure (reproducibility). The common recommendation to sample a field 20 times and mix all the samples into one composite, produces an accuracy of about ±17% for NO_3^-, assuming 80% precision. This is an acceptable level of accuracy and precision for fertilizer recommendation purposes in most cases.

In addition to adequate sample numbers, proper soil sampling techniques include:

- Where and how to sample each field (sampling plan)
- Proper equipment
- Proper sampling time
- Correct depth
- Proper sample handling

SOIL SAMPLING PLAN

Sample individual fields separately. The first step is to assess the field variability and identify representative areas. The type and level of variability can influence the choice of sampling plan.

Consider the four basic sampling plans:

1. Random soil sampling uses a random pattern across a field, generally avoiding unusual or problem areas such as hilltops and potholes. Bulk together 20 soil cores into one composite sample, air-dry then send to a soil test lab. This common method is adequate for smaller, relatively uniform fields.
2. Topographic sampling involves separating sets of samples based on topography. Identify the dominant topographic features such as hilltop, midslope and bottom slope, and take 20 core samples for each type. Bulk sub-samples from each type into one composite sample for each landscape type. Send several samples to the testing lab. This method can provide more meaningful results for variable fields, but at additional expense and labour. You must be willing and able to apply different fertilizer rates in the separate landscape areas.

3. Benchmark sampling expands on the topographic sampling concept by considering unique areas based on topography, soil texture and type and typical crop growth. Once these unique areas are identified, each year the samples are taken from the same spot within each benchmark area. There may be one or more benchmark sites within a field, depending on the variability. Each benchmark area then becomes a reference area on which fertilizer recommendations are based. The benchmark sampling area is much smaller than the whole field, and together with sampling in the same spot from year to year, this is assumed to reduce variability of the test results. If the benchmarks are carefully selected to represent the majority of the field, then good soil test results can be obtained.
4. Grid or systematic sampling follows an organized grid pattern, perhaps every 0.2 to 2 ha (0.5 to 5 acres). This method can reveal field nutrient variability and allow for variable rate fertilization and precision farming techniques. However, the sampling and analysis costs are not economic with current field crop prices on the prairies.

PROPER SOIL SAMPLING EQUIPMENT

Soil sampling to a 60 cm (2') depth can be done with a probe or auger. Do not use flight or screw sampling augers if samples need to be separated by depth since mixing will occur. If testing for micronutrients, ensure the sampling tool is chrome plated or stainless steel and rust-free. Do not bulk samples in metal pails. Clean plastic pails labelled for location and depth work well. Information sheets, sample bags and shipping boxes are available from soil testing labs and most fertilizer dealers.

PROPER SAMPLING DATE

The most accurate sampling time is just prior to seeding. However, this is practical because time is needed to purchase and perhaps place the fertilizer before seeding. Sampling is commonly done in the spring or fall. Spring sampling is done after the soil has thawed and is no longer saturated from snow melt. Fall sampling can begin once the soil has cooled to 5 to 7Â°C. This helps reduce nutrient content changes due to microbial activity.

The old assumption that N availability does not change over the winter has been proven wrong. Research in Alberta found that available N increased by 56 kg/ha (50 lb/ac) in stubble fields from fall to late winter while the soil was frozen. Summerfallow fields increased by 73 kg/ha (65 lb/ac). This overwinter gain in available N is apparently due to death of soil microbes and subsequent release of available N forms from their ruptured cells. However, the available N gained over the winter was temporary as it was mostly lost in the spring, probably through denitrification. The stubble fields lost 45 kg/ha (40 lb N/ac) in the spring while summerfallow fields lost 73 kg/ha (65 lb/ac). Therefore, the net change from late fall to late spring was minimal.

Late fall sampling tends to more accurately reflect spring NO_3^- (nitrate) contents than early fall sampling, especially for Black soils. Alberta research in the 1980s compared soil samples taken in the fall (early October and early November) to spring samples for 26 stubble fields. Early fall samples averaged 34 kg/ha (30 lb/ac) less nitrate N than spring samples, while late fall samples averaged only about 17 kg/ha (15 lb/ac) less. The early fall samples were also more variable in relation to spring samples. Overall, late fall samples more accurately predicted spring nitrate contents and grain yields than early fall. Spring samples were slightly better than late fall samples for predicting grain yield and N uptake.

In contrast, recent research in Manitoba measured very little change in soil nitrate levels in cereal stubble from early September to freeze-up. North and South Dakota extension soil scientists recommend early fall sampling in view of time constraints, but reduce the N recommendation by 0.2 kg (0.5 lb) for each sampling day prior to September 15.

Another experiment in central Alberta compared the effects of sample timing on phosphorus (P) soil test values. On average over 27 sites, the extractable P increased from 28 kg/ha (25 lb/ac) in early October to 49 kg/ha (44 lb/ac) by early November, and to 50 kg (45 lb) by spring (late April to early May). The relationship between early fall and spring P values was not close and, therefore, it was not possible to simply correct the early fall values. In contrast, research on a Brown soil in Saskatchewan over 24 years found both overwinter increases and decreases in soil P tests, but relatively few were significant. An experiment on irrigated alfalfa on a Dark Brown soil near Lethbridge, AB found significant overwinter increases in organic P. The conflicting results may be related to the differences in soil organic matter content and biological activity, and, therefore, potential for microbial changes to the plant available P pool.

In conclusion, early fall sampling can create higher than necessary fertilizer N and P recommendations due to an underestimation of spring nitrate N and available P, especially in the Black and Gray soil zones. Fertilizer response curves have been calibrated only against spring nutrient contents.

One disadvantage to late fall sampling after soil has cooled to 5 to 7°C is that fall fertilizer banding opportunities become more limited. By the time the samples are taken, dried, sent to the lab, analysed and results returned, the soil may have become frozen or covered with snow. On average over the prairies, soil cools by 1°C every five days in the fall.

CORRECT SAMPLING DEPTH

The appropriate sampling depth depends on the nutrients to be tested. For mobile nutrients such as N and sulphur (S), sampling to the 60 cm (2') depth is usually the most accurate according to research conducted in the 1960s. The fertilizer response database for N was developed with 0 to 60 cm (0 to 24") samples. However, recent research in Saskatchewan and North Dakota indicates that the 0 to 30 cm (0 to 12") depth may be more accurate for N than either 0 to 15 cm (0 to 6") or 60 cm samples. This recent research and the fact that sampling 60 cm is considerably more difficult, supports the 0 to 30 cm depth as a reasonable recommendation for these areas. In contrast, research in Manitoba has documented that the 0 to 15 cm depth is inferior. Separate samples for 0 to 15 cm, 15 to 30 cm and 30 to 60 cm was often recommended in the past, but was rarely done due to additional expense and time.

For immobile nutrients such as P and potassium (K) and most micronutrients, the ideal depth is 0 to 15 cm because the fertilizer response calibrations are based on that depth. If only the 0 to 30 cm depth is sampled, the soil testing labs must use a correction factor to estimate the value for the 0 to 15 cm depth. Since these nutrients are relatively immobile, they tend to remain at the fertilizer application depth. Therefore, the 0 to 30 cm depth may underestimate these nutrients and lead to high fertilizer recommendations.

PROPER SAMPLE HANDLING

Handle samples carefully to prevent accidental mixing and contamination. Mix each sample and spread on separate pieces of clean paper to dry at room temperature. Use a fan if required and avoid additional heat sources. Once dry, fill soil sample cartons or bags with about 0.5 kg (1 lb) of soil, and label with field number and depth.

SOIL SALINITY

All soils contain soluble salts (those which dissolve in water). When the levels are sufficient to harm plants, the soils are "saline."

Salinity is the result of excess groundwater moving downward and laterally through the soil dissolving and transporting soluble salts. Large areas in Canada's prairie regions contain soil materials relatively high in soluble salts. The redistribution of these soluble salts by groundwater movement causes some areas to become excessively saline. When drainage is good, the salts are washed down through the soil and out of the root zone. When drainage is poor, as in low, flat or depressional areas, or in areas where roads or other construction interferes with normal drainage, the water table rises. When the water table rises to within 1 m (3') of the surface, water and salts can rise to the surface by capillary action. The water evaporates and the salts accumulate at the surface. Salinity also occurs in saline seeps on hill slopes, often part way up the slope on very long and high hills. Rainfall on the upslope part of the hill (recharge area) moves down through the soil, picking up salts on the way. The water not used by crops moves down until it reaches an impermeable layer, which impedes its progress. The water then flows laterally (seeps) until it reaches a position lower down the slope (discharge area) where the water table is closer to the surface. There, seep water causes the water table to rise allowing the salt laden water to move up to the soil surface by capillary action.

The seep phenomenon can take place over relatively short distances within the same field or over distances of several miles. Salinity levels vary widely across a saline seep. Salinity also varies from spring to fall. Salinity usually appears on the soil surface just after spring thaw.The dominant salts in prairie "saline seeps" are calcium (Ca), magnesium (Mg), and sodium (Na) cations and sulphate (SO_4) anions. If Na levels are high or not balanced with the Ca and Mg, soil tilth can also be affected. The positively charged Na cations attach to the negatively charged clay particles in the soil, causing the soil to be sticky when wet and hard and impermeable when dry. Saline soils can be recognized by spotty growth of crops or by white crusts of salt, which accumulate on the soil surface usually in low-lying areas. Streaks of salt may be present in the soil even though white crusts may not appear on the surface. In many saline soils, it is not possible to see the salts and a laboratory soil analysis must be used to confirm their presence. Plants may develop a blue-green tinge. Salttolerant weeds such as Russian thistle, kochia, wild barley and goosefoot species are commonly found in areas of high salt concentration. The process of salt buildup is reduced when plants intercept the upward flow of saline groundwater and reduce the amount that reaches the soil surface. When plants are not present, almost all of the water loss takes place by evaporation at the soil surface.

Summerfallow is a major contributor to salinity as it encourages a build-up of a water table. Water not used by crops accumulates in the subsoil eventually coming to the surface laden with salts. These salts interfere with seed germination and crop establishment. All irrigation water contains some salt. Over an extended period of irrigation where soil drainage is inadequate, salts accumulate in the soil and a salinity problem may develop. Water used in the major irrigation areas of Alberta and Saskatchewan is of high quality with little risk of salt buildup. Seepage or high water tables resulting from overirrigation cause most salinity under irrigation.

EFFECTS OF SALINITY ON CANOLA GROWTH AND YIELD

Plant species vary in their tolerance to salt affected soils. Some plants will tolerate high levels of salinity while others can tolerate little or no salinity.

The relative growth of plants in the presence of salinity is called their salt tolerance. Salttolerant plants avoid toxicity by the sequestration of the Na+ and Cl- ions into either vacuoles or roots by a salt exclusion mechanism. Salt tolerances are usually given in terms of stage of plant growth over a range of electrical conductivity (EC) levels. The degree of salinity or total soluble salt concentration in a soil is routinely measured by soil testing laboratories with a conductivity test. Electrical conductivity (EC) is the ability of a solution to transmit an electrical current. An electrical current is passed through a soil sample and measured. The units of conductivity are usually given in deciSiemens per metre (dS/m). Table categorizes salinity into general ranges from non-saline to very severely saline. These values are used for plant selection for saline soils.

Excess soluble salts cause osmotic stress and ion toxicity in plant cells. Plants need both the water and the nutrients dissolved in it for proper growth. The sap in plant roots contains salt that attracts water into the plant via osmotic pressure. Dissolved salts in the soil increase the osmotic pressure of the soil solution. This decreases the rate at which water from the soil will enter the roots. If the soil solution becomes too concentrated, plants will slowly dehydrate, lose turgor and starve, even though the supply of water and nutrients in the soil may be quite high.

If the salt content of the soil water becomes too saturated, water may actually be withdrawn from the roots. Osmotic stress also results from desiccation and, therefore, is a common component of both drought and salt stress. High concentration of certain salts in the soil may also be toxic because some plants may absorb an excess amount of the salt, reducing growth or causing death.

Canola is considered moderately tolerant to salt and sodium, and can tolerate moderate salinity up to levels of 5 to 6 dS/m. In a salinity study by the Agriculture and Agri- Food Canada Swift Current, SK Research Centre increasing levels of soil salinity reduced germination, emergence and emergence rate. This greenhouse study showed that the per cent emergence and plant survival was not significantly affected until salinity exceeded 5.6 dS/m.

Above this value there was an increase in number of days from seeding to initial emergence. With severe salinity, after emergence began, there was a rapid number of new plants but over time the percentage that survived dropped off. Salinity under both moderate and severe conditions reduces average plant height, shoot and root biomass, number of leaves, leaf area, evapotranspiration, and crop yield at harvest. For example, the average harvest height in the above study averaged 167, 132 and 84 cm (66, 52 and 33"), respectively for the 0.6, 5.6 and 12.4 dS/m treatments. Root growth is also hindered at salinity above 6 dS/m.

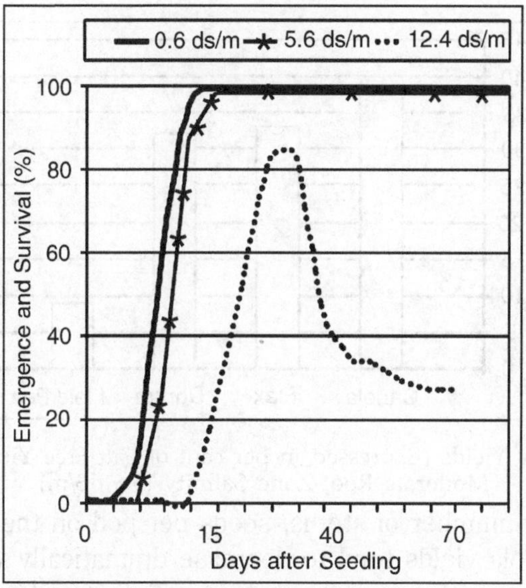

Fig. Effects of Salinity (dS/m) on per cent Emergence, Days to Emerge and Plant Survival

At 5.6 dS/m, canola yields in this study were reduced by 60% and with severe salinity, canola yields were reduced by 80%. However, several other research studies have shown there was little effect on canola yields with salinity levels up to 10 dS/m.

Use soil tests to determine the extent of a salinity problem. Sample both affected and non-affected areas of the field. Perform analyses for electrical conductivity (EC), pH, cation base saturation, and calcium, magnesium, sodium and organic matter content. Soil tests can also identify possible future salinity problems.

Another type of soil problem occurs when sodium levels are high in relation to calcium and magnesium in the soil.

These soils are very sticky and slippery when wet, and very hard, cloddy and prone to crusting when dry. The sodium adsorption ratio (SAR) is the ratio of sodium to the beneficial soil structural cations, calcium and magnesium.

When the SAR value exceeds 13, the soil is "sodic." If the SAR exceeds 13 and the EC is greater than 4, it is considered a "salinesodic" soil. Use a soil test to determine the SAR of soils.

Sodium, even with no soil crusting, has been shown by an Argentina study to have a direct effect on canola emergence with a slight reduction in emergence at SAR values of 20 growing to a 60% reduction at SAR values above 34. At SAR values greater than 20, main stem growth and yield decreased slightly but an increase in the number of secondary branches compensated for this reduction.

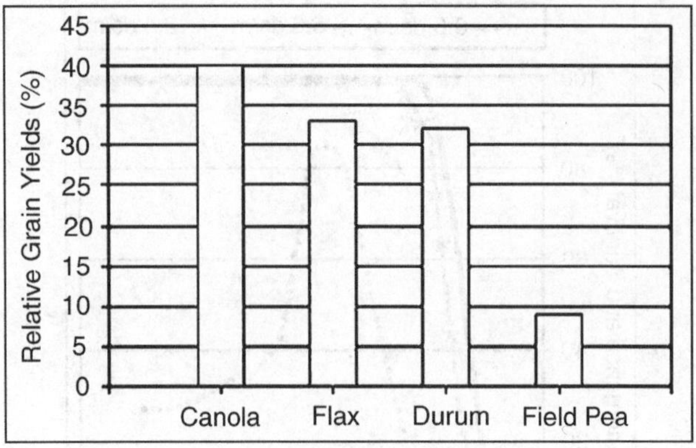

Fig. Canola Yields (Expressed in per cent of Salt-Free Yields) under Moderate Root-Zone Salinity (5.6 dS/m)

Reductions in number of stems, seeds per pod on the main stem, pods per stem and canola yields tend to decrease dramatically when SAR values exceed 34. However, the most important effect of soil sodium is through increased crusting that reduces emergence. The amount of sodium required to affect emergence is much higher than that necessary for clay dispersion and crusting. The Argentina study showed SAR values of eight and even lower can encourage clay dispersion and crusting when raindrops impact the soil surface.

CONSISTENCY OF SOIL TEST LAB RECOMMENDATIONS

Soil sampling, analysis and interpretation is not an exact science. However, reasonable precision and accuracy is needed in order to make costly fertilizer decisions. Over the years, various growers, agencies or companies have sent duplicate samples to different labs to compare their analysis and recommendations. Unfortunately, widely differing results and recommendations have occurred, causing growers to question the credibility of the labs or the soil testing practice.

Two fundamental reasons contribute to differences in results. First, labs may be using different analytical procedures to measure soil nutrient content. For example, there are several methods to extract soil phosphorus. Or the technique may differ slightly when using a particular method. Over time, labs are harmonizing their test methods and in time this problem may disappear.Variations in soil test recommendations arise mainly due to the differences in each labs interpretation. The recommended fertilizer amounts at the same soil test level can vary significantly from lab to lab.

This may be due to using:

- Different critical (deficient) soil levels being used
- Regional fertilizer response (calibration) data but modifying recommendations to fit a particular philosophy of fertilizer use or economic payback

- Recommendations from other regions or countries
- A unique system of fertilizer recommendation not based on regional calibration data or economics

Although consistency among labs has been improving since the first comparison studies in the 1980s, further improvements are needed. Calibration data must continue to be collected to account for changes in fertilizer application techniques and changes in other agronomic practices.

Overall, soil testing is a useful agronomic practice. Use labs that base fertilizer recommendations on economics using regional calibration data. Be prepared to question unusual recommendations based on experience and the local knowledge of qualified agronomists. Keep in mind that the accuracy of fertilizer recommendations will always be limited by sampling challenges and the inability to predict the weather of the upcoming growing season.

HOW TO DIAGNOSE NUTRIENT DEFICIENCY SYMPTOMS

In moderate to severe nutrient deficiencies, visible symptoms can indicate the specific nutrient that is lacking. Nutrients that are only slightly limiting often do not show visible symptoms, a situation that has been termed hidden hunger.

A systematic diagnosis of visible symptoms is needed to correctly identify limiting nutrients. Symptoms usually appear on either old or young leaves depending on the mobility of the nutrient in question. Chlorosis (loss of green colour, yellowing) and necrosis (death of plant tissue, often leading to white or brown colour) are important visible symptoms. Diagnosis under field situations can be complicated by high field variability, multiple deficiencies, and other causes such as weather, pests and herbicide injury. For example, a sulphur deficiency can easily be confused with Group 2 herbicide injury due to similar symptoms.

PLANT AND TISSUE TESTING

Crop nutritional status also can be assessed by plant and tissue analyses. These methods can supplement, but not replace, soil testing. Plant and tissue analyses measure the nutrient content of above ground plant parts during growth. The values are compared to established ranges for inadequate, adequate and excess levels.

Plant tissue testing is suitable for diagnosing crop problems that may be nutritionally related and to identify any nutrients that may be limiting yields. Plant analysis can determine if the fertilizer rate and method of application were adequate to meet crop needs. The disadvantage of tissue sampling is timing after tissue samples are taken and analysed, it may be too late to correct deficiencies in the current crop. No reliable interpretative criteria exist for nutrient ranges in seedling canola. Also, nutrient contents usually differ greatly between different plant parts and ages. Therefore, the proper part must be sampled at the proper growth stage. An adequate sample will contain 50 to 80 plants, depending on the nutrients to be tested and plant part/age. Avoid unusual, dead or stressed plants, as well as those covered with soil or recent sprays. Cut samples with a clean, rust-free knife or scissors. Dry the plant samples on clean paper or plastic at room temperature (do not oven dry). After drying, keep the samples in a paper bag.

3

Solid Supports for the Synthesis of Peptides

The revolutionary principle behind solid phase peptide synthesis (SPPS) as conceived by Merrifield in 1959 is that if the peptide is bound to an insoluble support then any unreacted reagents left at the end of any synthetic step can be removed by a simple wash procedure, greatly decreasing the time required for synthesis. What is more, the arrangement is amenable to automation. This is only valid, however, if the individual synthetic steps occur with essentially quantitative yields. This latter requirement entails rigorous testing of the chemistry and tends to result in the use of a limited range of tried and tested methodologies.

The two most commonly used schemes are the original Merrifield method and that proposed by Sheppard in 1975. While the protection schemes used are different, they have the same basic concepts, a resin support, the use of an excess of reagents and building the peptide in a C→N terminal direction.

The resin supports used are designed to swell in commonly used solvents, expanding to many times their original size. Thus, the reactions occur not on the surface of a rigid particle, but within the support in a solvated gel which permits easy access to the growing peptide chain.

Reagents are used in excess, this allows the reactions to proceed to completion in the minimum time and results in faster synthesis of peptides with high purity. The couplings proceed in a C→N terminal direction to allow the use of racemisation limiting amine protection for the activated species.

MERRIFIELD SYNTHESIS

This methodology is characterised by the use of *tert*-butyl based temporary a-amino protection and benzyl, or substituted benzyl, groups for permanent side chain protection. There are over one hundred different substituted resins suitable for peptide synthesis generally based on polystyrene and polyethylene glycol. These resins allow introduction of an amino acid through either substitution, condensation or addition reactions.

The traditional resin used for Merrifield synthesis was a chloromethylphenyl substituted resin. The first amino acid was attached to the resin through substitution of the chloride by the caesium salt of the BOC-amino acid, generating an equivalent to a benzyl ester.

Deprotection of the temporary BOC group uses a 20-50% solution of trifluoroacetic acid (TFA) in dichloromethane and has to be followed by neutralisation of the resulting ammonium salt with a hindered tertiary base. Final cleavage from the resin as well as deprotection of benzyl based side chain protecting groups is achieved using strong acids, usually liquid hydrogen fluoride or trifluoromethane sulphonic acid. Such procedures require specialised apparatus and the highly acidic conditions catalyse several possible rearrangements.

FMOC POLYAMIDE SYNTHESIS

The fundamental differences between the Fmoc polyamide strategy when compared to the Merrifield approach are that the reactions are carried out under continuous flow and that the conditions for a-amino deprotection and cleavage from the resin are far more mild.

This arises from the adoption of the base labile Fmoc protecting group for a-amino protection. The side chains are generally protected with *tert*-butyl based groups which, in common with the linkage to the resin, can be cleaved by TFA in the presence of scavengers.

Traditionally, resins with 4 hydroxymethylphenoxy substitution were used. These allowed were esterified with the anhydride of the first amino acid. As a result of the mesomerically electron donating *para* oxygen atom stabilising the resultant carbocation, cleavage of the peptide from the resin occurs under more mild acid conditions, typically using trifluoroacetic acid with scavengers.

The use of continuous flow means that the reagents are passed through a reaction chamber containing the resin supported peptide. Having passed through this chamber, the reagents can be recirculated back into the chamber again or taken to a waste collection bottle.

This allows the resin to be washed clean of excess reagents and unwanted reaction products, which in turn helps drive deprotection steps to completion following Le Chatelier's principle. In addition, the solution can be passed through a u.v. detector and monitored at a suitable wavelength for the Fmoc chromophore.

A typical cycle consists of:

- Deprotection of the preceding residue with piperidine.
- Wash to remove any remaining reagents from 1.
- Acylation in a recirculatory mode.
- Wash to remove excess reagents.

A typical uv trace indicates these steps. In this way, the cycle can be qualitatively monitored. Such monitoring can be used in conjunction with automation. For example, a slow deprotection step suggests inaccessibility of the peptide amino terminal, indicating that the next coupling step may require a longer acylation time. If this is detected then the cycle can be interrupted for later manual intervention.

Since the turn of the century, there has been a constant drive to develop new and improved strategies for peptide synthesis. The areas of protection, deprotection, activation and coupling have all received attention. The field of solid phase peptide synthesis has rapidly grown since its inception by Merrifield but is dominated by use of BOC/benzyl and Fmoc/*tert*-butyl protection schemes.

SOLUTION PHASE PEPTIDE SYNTHESIS

Care has to be taken in selection of protecting groups so that they can be removed with appropriate selectivity. Side chain and carboxyl protecting groups need to be selected so as not to be labile under the conditions required to deprotect the amino group. If a long peptide is to be made then it is common to synthesise it in smaller sections, which are later coupled to form the overall peptide-a process called fragment condensation. If a fragment condensation technique is being adopted then the carboxyl terming must also be deprotected without loss of side chain protection.

There are two ways of achieving this selective deprotection:

1. Chose protecting groups that are deprotected with completely different reagents (refered to as orthogonal protection) *e.g. tert butyl* (acid), fluorenyl methyl (base), benzyl (catalytic hydrogenolysis).
2. Chose protecting groups that are deprotected with the same type of reagent but under different conditions. *e.g. tert* butyl and benzyl which require increasingly strong acids for deprotection.

SOIL ORGANIC MATTER

On the basis of organic matter content, soils are characterized as mineral or organic. Mineral soils form most of the world's cultivated land and may contain from a trace to 30 per cent organic matter. Organic soils are naturally rich in organic matter principally for climatic reasons. Although they contain more than 30 per cent organic matter, it is precisely for this reason that they are not vital cropping soils. This soils bulletin concentrates on the organic matter dynamics of cropping soils. In brief, it discusses circumstances that deplete organic matter and the negative outcomes of this. The bulletin then moves on to more proactive solutions. It reviews a "basket" of practices in order to show how they can increase organic matter content and discusses the land and cropping benefits that then accrue. Soil organic matter is any material produced originally by living organisms (plant or animal) that is returned to the soil and goes through the decomposition process. At any given time, it consists of a range of materials from the intact original tissues of plants and animals to the substantially decomposed mixture of materials known as humus.

Fig. Plate Crop Residues Added to The Soil are Decomposed by Soil Macrofauna and Micro-Organisms, Increasing The Organic Matter Content of The Soil.

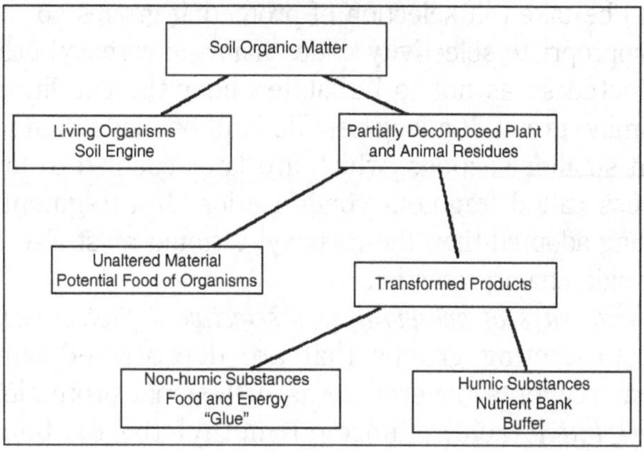

Fig. Components of Soil Organic Matter and Their Functions

Most soil organic matter originates from plant tissue. Plant residues contain 60-90 per cent moisture. The remaining dry matter consists of carbon (C), oxygen, hydrogen (H) and small amounts of sulphur (S), nitrogen (N), phosphorus (P), potassium (K), calcium (Ca) and magnesium (Mg). Although present in small amounts, these nutrients are very important from the viewpoint of soil fertility management.

Soil organic matter consists of a variety of components. These include, in varying proportions and many intermediate stages, an active organic fraction including microorganisms (10-40 per cent), and resistant or stable organic matter (40-60 per cent), also referred to as humus.

Forms and classification of soil organic matter have been described by Tate and Theng. For practical purposes, organic matter may be divided into aboveground and belowground fractions. Aboveground organic matter comprises plant residues and animal residues; belowground organic matter consists of living soil fauna and microflora, partially decomposed plant and animal residues, and humic substances. The C:N ratio is also used to indicate the type of material and ease of decomposition; hard woody materials with a high C:N ratio being more resilient than soft leafy materials with a low C:N ratio.

Although soil organic matter can be partitioned conveniently into different fractions, these do not represent static end products. Instead, the amounts present reflect a dynamic equilibrium.

The total amount and partitioning of organic matter in the soil is influenced by soil properties and by the quantity of annual inputs of plant and animal residues to the ecosystem. For example, in a given soil ecosystem, the rate of decomposition and accumulation of soil organic matter is determined by such soil properties as texture, pH, temperature, moisture, aeration, clay mineralogy and soil biological activities. A complication is that soil organic matter in turn influences or modifies many of these same soil properties.

Organic matter existing on the soil surface as raw plant residues helps protect the soil from the effect of rainfall, wind and sun. Removal, incorporation or burning of residues exposes the soil to negative climatic impacts, and removal or burning deprives the soil organisms of their primary energy source. Organic matter within the soil serves several functions.

From a practical agricultural standpoint, it is important for two main reasons:

1 As a "revolving nutrient fund"; and
2. As an agent to improve soil structure, maintain tilth and minimize erosion.

As a revolving nutrient fund, organic matter serves two main functions:

1. As soil organic matter is derived mainly from plant residues, it contains all of the essential plant nutrients. Therefore, accumulated organic matter is a storehouse of plant nutrients.
2. The stable organic fraction (humus) adsorbs and holds nutrients in a plant-available form.

Organic matter releases nutrients in a plant-available form upon decomposition. In order to maintain this nutrient cycling system, the rate of organic matter addition from crop residues, manure and any other sources must equal the rate of decomposition, and take into account the rate of uptake by plants and losses by leaching and erosion.

Where the rate of addition is less than the rate of decomposition, soil organic matter declines. Conversely, where the rate of addition is higher than the rate of decomposition, soil organic matter increases. The term steady state describes a condition where the rate of addition is equal to the rate of decomposition. In terms of improving soil structure, the active and some of the resistant soil organic components, together with micro-organisms (especially fungi), are involved in binding soil particles into larger aggregates. Aggregation is important for good soil structure, aeration, water infiltration and resistance to erosion and crusting.

Traditionally, soil aggregation has been linked with either total C or organic C levels. More recently, techniques have developed to fractionate C on the basis of lability (ease of oxidation), recognizing that these subpools of C may have greater effect on soil physical stability and be more sensitive indicators than total C values of carbon dynamics in agricultural systems. The labile carbon fraction has been shown to be an indicator of key soil chemical and physical properties.

For example, this fraction has been shown to be the primary factor controlling aggregate breakdown in Ferrosols (non-cracking red clays), measured by the percentage of aggregates measuring less than 0.125 mm in the surface crust after simulated rain in the laboratory.

The resistant or stable fraction of soil organic matter contributes mainly to nutrient holding capacity (cation exchange capacity [CEC]) and soil colour. This fraction of organic matter decomposes very slowly. Therefore, it has less influence on soil fertility than the active organic fraction.

WHAT IS ORGANIC MATTER

Soil organic matter consists of a variety of components. *These include, in varying proportions and many Intermediate stages:*

- Raw plant residues and microorganisms (1 to 10 per cent)
- "Active" organic traction (10 to 40 per cent)
- Resistant or stable organic matter (40 to 60 per cent) also referred to as humus.

Raw plant residues, on the surface, help reduce surface wind speed and water run-off. Removal, incorporation or burning of residues predisposes the soil to serious erosion.

The "active" and some of the resistant soil organic components, together with microorganisms (especially fungi) are involved in binding small soil particles into larger aggregates. Aggregation is important for good soil structure, aeration, water infiltration and resistance to erosion and crusting. The resistant or stable fraction of soil organic matter contributes mainly to nutrient holding capacity (cation exchange capacity) and soil colour.

This fraction of organic matter decomposes very slowly and therefore has less influence on soil fertility than the "active" organic fraction. Organic matter in soil serves several functions.

From a practical agricultural standpoint, it is important for two main reasons. First as a "revolving nutrient bank account"; and second, as an agent to improve soil structure, maintain tilth, and minimize erosion. *As a revolving nutrient bank account, organic matter serves two main functions:*

1. Since soil organic matter is derived mainly from plant residues, it contains all of the essential plant nutrients. Accumulated organic matter, therefore, is a storehouse of plant nutrients. Upon decomposition, the nutrients are released in a plant-available form.
2. The stable organic fraction (humus) adsorbs and holds nutrients in a plant available form.

Organic matter does not add any "new' plant nutrients but releases nutrients in a plant available form through the process of decomposition. In order to maintain this nutrient cycling system, the rate of addition from crop residues and manure must equal the rate of decomposition. If the rate of addition is less than the rate of decomposition, soil organic matter will decline and, conversely if the rate of addition is greater than the rate of decomposition, soil organic matter will increase.

The term steady state has been used to describe a condition where the rate of addition is equal to the rate of decomposition. Fertilizer can contribute to the maintenance of this revolving nutrient bank account by increasing crop yields and consequently the amount of residues returned to the soil.

ORGANIC MATTER IN VIRGIN AND CULTIVATED SOILS

Soils in Alberta are divided into soil groups (zones) based on the amount of organic matter they contain. They occur in geographic zones from the southeast to the northwest and are identified as the Brown, Dark Brown, and Black Chernozemic (prairie) soils. The Brown soils have the least amount of organic matter because of the relatively small inputs of plant residues contributed by the short grass prairie vegetation under which these soils developed. Black soils developed under cooler and wetter conditions which allowed for more grass growth and thus a greater accumulation of organic matter. Further north and west, trees became the dominant vegetation. Soils influenced by forest vegetation for a moderate length of time constitute the Dark Gray or transitional soils. Where the forest cover was established for a longer period, Luvisolic (forest) soils developed. Organic (peat) soils. occur in low lying areas throughout the Black, Dark Gray and Gray soil zones. These soils are saturated with water for much or all of the year thereby reducing the rate of organic matter decomposition. The amount of soil organic matter characteristic of virgin and cultivated soils in the various zones is shown in Table 1. Cultivation generally has resulted in a 30 to 50 per cent loss of organic matter.

Table. Organic Matter in Native and Cultivated Soils (per cent)

Soil zone	Virgin	Cultivated
Brown	3-4	2-3
Dark Brown	4-5	3-4

Black	6-10	4-6
Dark Gray	4-5	2-3
Gray	1-2	1-2

Before our soils were cultivated, they had achieved a "steady state". In most of our prairie soils, the increased rate of decomposition associated with cultivation, combined with the low rates of crop residue addition associated with crop-fallow rotations has caused a fairly rapid decline in soil organic matter. The rate of decline decreases with time as the amount of total soil organic matter decreases and particularly as the "active" organic fraction is depleted. Cultivation of soils that are naturally high in organic matter will usually result in a decrease of organic matter. In the case of Luvisolic soils, their poor physical properties and low fertility have encouraged the use of forages, fertilizers, manure and judicious tillage. Such management practices have resulted in an increase in soil organic matter on Luvisolic soils, whereas excessive tillage, fallowing and minimal fertilization have lead to further depletion of the soil organic matter.

EFFECTS OF ORGANIC MATTER DECLINE

As stated in the introduction, soil degradation is becoming a major concern in Canada. Loss of organic matter is often identified as one of the main factors contributing to declining soil productivity, but it is misleading to equate a loss in soil organic matter with a loss in soil productivity. Soil organic matter contributes to soil productivity in several ways, but there is no direct quantitative relationship between soil productivity and total soil organic matter. In fact, it has been the decline in organic matter that has contributed to the productivity of the crop-fallow system.

This decline in organic matter has resulted in the release of large amounts of plant nutrients, particularly nitrogen. For example, a decrease in soil organic matter of 2 per cent releases about 2,400 lb/ac of nitrogen. If this decline occurred over a 60 year period, an average of 40 lb/ac/yr of plant-available nitrogen has come from the soil organic matter. We therefore view prairie soils which had relatively high levels of organic matter as being nitrogen fertile, but this fertility could only be attained under a management system that allowed for organic matter to decline.

Frequent fallowing has been a major factor contributing to this decline. Insofar as organic matter contributes to improved soil physical properties (*e.g.*, tilth, aggregation, moisture holding capacity and resistance to erosion) increasing soil organic matter will generally result in increased soil productivity. But on many soils, suitable soil physical properties occur at relatively low levels of organic matter (2-4 per cent). A level of organic matter higher than required to produce suitable physical properties is beneficial in that the soil has a greater buffering and nutrient holding capacity, but it does not contribute directly to soil productivity. If soils are managed so organic matter is not declining (steady-state), soils higher in organic matter (*e.g.*, 8 per cent) are not inherently more productive or fertile than those that have less organic matter (*e.g.*, 5 per cent). To equate the ability to supply nutrients with total soil organic matter is not valid. The "active" fraction of organic matter is a more reliable indicator of soil fertility than is total soil organic matter. In cultivated soil, the "active" fraction is influence mainly by previous management. Soil organic matter cannot be increased quickly even when management practices that conserve soil organic matter are adopted. The increased addition of organic matter associated with continuous cropping, and the production of higher crop yields, are accompanied by an increase in the rate of decomposition. Moreover, only a small fraction of crop residues added to soil remains as soil organic matter. After an extended period of time, the return of all crop residues and the use of forages in rotations with cereals and oilseeds may significantly increase soil organic matter, particularly, the "active" fraction.

MANAGING SOIL ORGANIC MATTER

There have been vast changes in the nature of agricultural production. In the past, farms were small, and much of what was produced was consumed on the farm. This system allowed for the limited removal of soil nutrients since there was an opportunity to return most of the nutrients back to the land.

The advent of the internal combustion engine, migration from rural to urban communities, increasing farm size and specialization in production have resulted in a system of production where there is greater removal of plant nutrients from the soil and less opportunity for nutrient cycling. Maintenance of organic matter for the sake of maintenance alone is not a practical approach to farming. It is more realistic to use a management system that will give sustained profitable production. The greatest source of soil organic matter is the residue contributed by current crops. Consequently, crop yield and type, method of handling residues and frequency of fallow are all important factors. Ultimately, soil organic matter must be maintained at a level necessary to maintain soil tilth.

Summerfallow

Summerfallowing accelerates the loss of organic matter. Aeration of the soil associated with tillage, and the increase in soil temperature and moisture results in increased organic matter decomposition. Since little In the way of residues are added to the soil, a net loss of organic matter occurs. Research has shown that as the frequency of fallow increases, the amount of soil organic matter decreases.

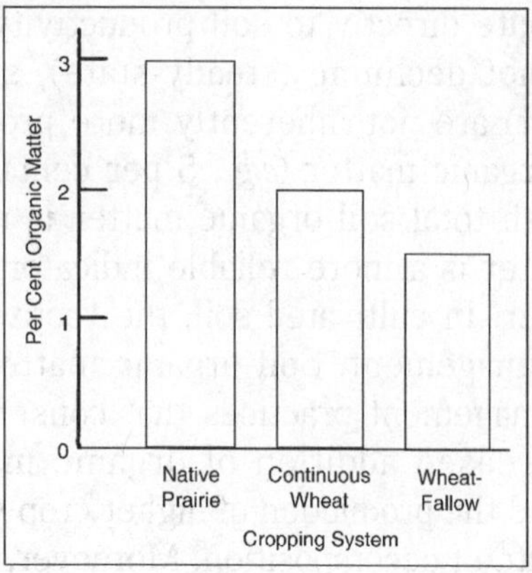

Fig. Effect of Frequency of Fallow on Per Cent Organic Matter.

Summer fallowing for moisture conservation may be a necessary practice in the Brown and Dark Brown soil zones. However, it must be questioned in the Black and the Gray soil zones. Periodic fallowing may be acceptable in the higher rainfall regions for control of persistent perennial weeds and volunteer grains in pedigreed seed production. In the crop-fallow system common to the prairie region, the nitrogen removed has far exceeded that gained from crop residues, manure, legumes and fertilizer. The large reserves of nitrogen present in the organic matter of our prairie soils have been the major source of nitrogen in this cropping system. Continued reliance on soil organic matter reserves to supply the nitrogen requirements of crops will ultimately lead to a decline in soil productivity, and increased soil erosion. When a change from a cropping system involving fallow to continuous cereal grain production, the nitrogen requirement increases. The nitrogen requirement is greatest in the first few years of continuous cropping as the nutrient cycling process adjusts to the new cropping system.

Crop Rotations

The value of forage crops in rotations with cereals and oilseeds has long been recognized, especially in the Luvisolic soils. The Breton Plots compared a two-year fallow-wheat rotation with a five-year rotation involving wheat, oats and barley followed by two years of hay production. In research done by the University of Manitoba, the effects of various cultural practices on the level of organic matter are compared. It is interesting to note that the highest level of soil organic matter was maintained under continuous cropping.

The beneficial effects of perennial forages are the result of:

- A more extensive root system and crop aftermath contributing more organic matter to the soil,
- The fibrous nature of the root system of perennial grasses. These are particularly effective as a binding agent in soil aggregation,
- Nitrogen fertility enhancement by the growth of legumes,

- Increased permeability of dense subsoils because of the deep penetrating tap roots of perennial legumes, especially alfalfa.
- A reduced rate of organic matter decomposition in the absence of tillage.

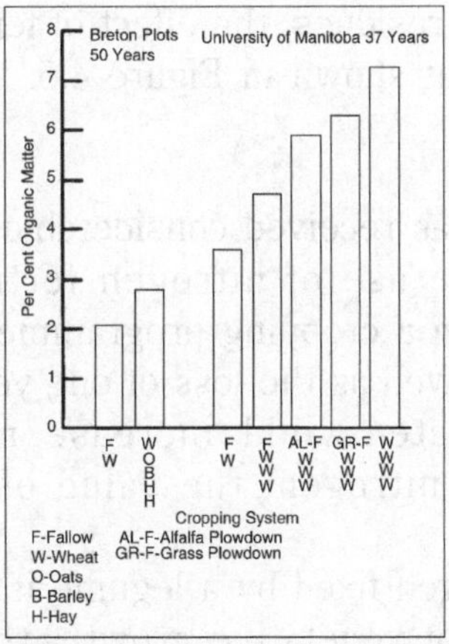

Fig. Effect of Rotation on Soil Organic Matter

In the brown and dark brown soil zones, perennial forages grown for forage production or as a plowdown crop may jeopardize subsequent cereal crops because they deplete soil moisture reserves.

Fertilization

Fertilizers will generally increase soil organic matter because the increased crop growth returns larger amounts of residues to the soil. Data obtained from the Breton Plots is summarized in Figure 4.4. The increase in organic matter is less than what might be expected with current farming practices since all the straw had been removed from the plots.

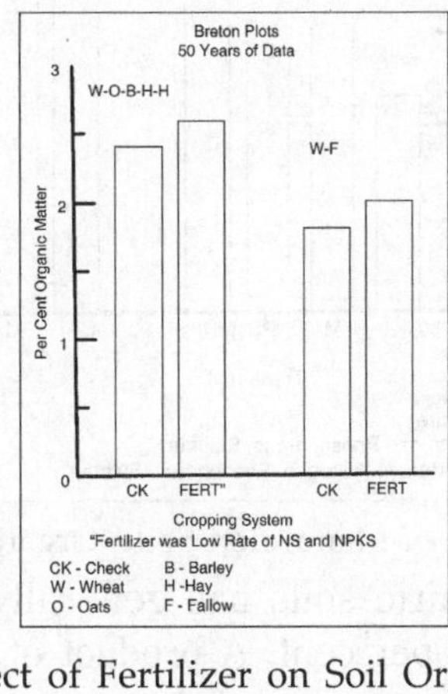

Fig. 4.5 Effect of Fertilizer on Soil Organic Matter.

To determine the fertilizer effect, two fertilizer treatments were averaged, one involving a low rate of nitrogen and sulphur and another involving a low rate of nitrogen, phosphorus, potassium and sulphur. One would expect that with higher rates of fertilization, higher yielding varieties, and the return of all crop residues, the effect of fertilizers on organic matter would be greater than that shown in Figure 4.5.

Plowdown

Legume plowdown has received considerable attention in recent years as an alternative to the use of nitrogen fertilizers. However, when considering this option in a cropping programme, the amount of nitrogen added by the legume, as well as the loss of one year of production, the cost of seed and the expected yield increase must be kept in mind. Strictly as a source of nitrogen, the value of a legume plowdown is questionable.

The amount of nitrogen fixed by a legume is dependent upon the type of legume, the amount of vegetative growth, the nature of the soil and environmental conditions. As a source of organic matter, legume plowdown is valuable, however, perennial forage is more effective than legume plowdown for increasing soil organic matter. Nitrate nitrogen which accumulates following legume plowdown is subject to loss, particularly in wet, poorly drained soils. To minimize this, legumes should be plowed down in the fall rather than mid-summer to reduce nitrate accumulation and subsequent loss.

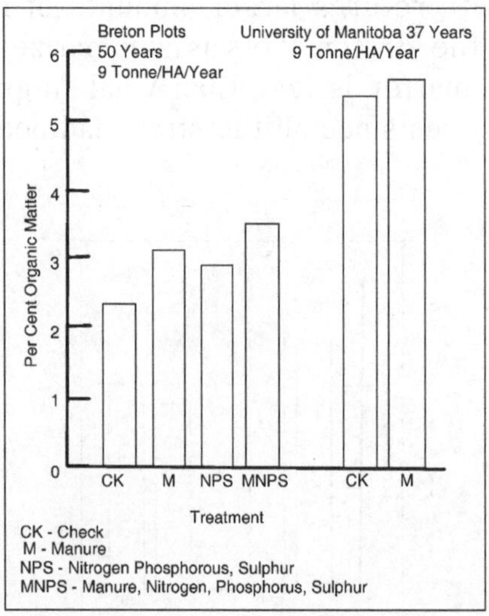

Fig. Effect of Manure on Soild Organic Matter.

The cultivation of prairie soils has generally resulted in a decline in organic matter of 30 to 50 per cent. A product of this decline has been the release of large amounts of plant nutrients, particularly nitrogen. Crop rotations with a high frequency of summerfallow have relied on the nitrogen released from soil organic matter to supply crop requirements. More frequent or continuous cropping, less frequent tillage, the production of high yields and the return of crop residues will help to maintain soil organic matter at a satisfactory level. Perennial forages are effective for maintaining or increasing soil organic matter.

ORGANIC MATTER DECOMPOSITION AND THE SOIL FOOD WEB

When plant residues are returned to the soil, various organic compounds undergo decomposition. Decomposition is a biological process that includes the physical breakdown and biochemical transformation of complex organic molecules of dead material into simpler organic and inorganic molecules.

The continual addition of decaying plant residues to the soil surface contributes to the biological activity and the carbon cycling process in the soil. Breakdown of soil organic matter and root growth and decay also contribute to these processes. Carbon cycling is the continuous transformation of organic and inorganic carbon compounds by plants and micro- and macro-organisms between the soil, plants and the atmosphere.

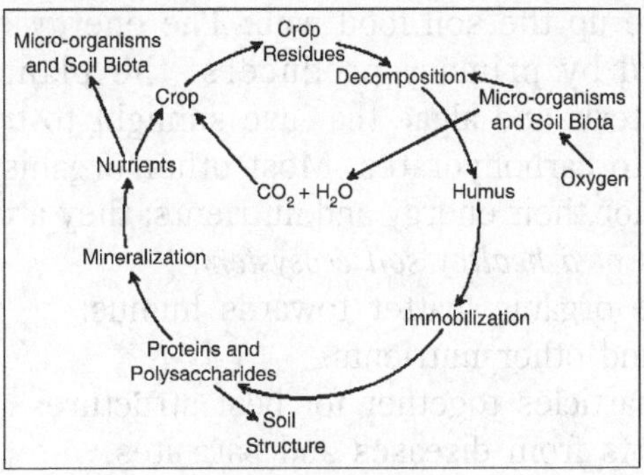

Fig. Carbon Cycle

Decomposition of organic matter is largely a biological process that occurs naturally. Its speed is determined by three major factors: soil organisms, the physical environment and the quality of the organic matter. In the decomposition process, different products are released: carbon dioxide (CO_2), energy, water, plant nutrients and re-synthesized organic carbon compounds. Successive decomposition of dead material and modified organic matter results in the formation of a more complex organic matter called humus.

This process is called humification. Humus affects soil properties. As it slowly decomposes, it colours the soil darker; increases soil aggregation and aggregate stability; increases the CEC (the ability to attract and retain nutrients); and contributes N, P and other nutrients.

Soil organisms, including micro-organisms, use soil organic matter as food. As they break down the organic matter, any excess nutrients (N, P and S) are released into the soil in forms that plants can use. This release process is called mineralization. The waste products produced by micro-organisms are also soil organic matter. This waste material is less decomposable than the original plant and animal material, but it can be used by a large number of organisms.

By breaking down carbon structures and rebuilding new ones or storing the C into their own biomass, soil biota plays the most important role in nutrient cycling processes and, thus, in the ability of a soil to provide the crop with sufficient nutrients to harvest a healthy product. The organic matter content, especially the more stable humus, increases the capacity to store water and store (sequester) C from the atmosphere.

THE SOIL FOOD WEB

The soil ecosystem can be defined as an interdependent life-support system composed of air, water, minerals, organic matter, and macro- and micro-organisms, all of which function together and interact closely.

The organisms and their interactions enhance many soil ecosystem functions and make up the soil food web. The energy needed for all food webs is generated by primary producers: the plants, lichens, moss, photosynthetic bacteria and algae that use sunlight to transform CO_2 from the atmosphere into carbohydrates. Most other organisms depend on the primary producers for their energy and nutrients; they are called consumers.

Some functions of a healthy soil ecosystem:

- Decompose organic matter towards humus.
- Retain N and other nutrients.
- Glue soil particles together for best structure.
- Protect roots from diseases and parasites.
- Make retained nutrients available to the plant.
- Produce hormones that help plants grow.
- Retain water.

Soil life plays a major role in many natural processes that determine nutrient and water availability for agricultural productivity. The primary activities of all living organisms are growing and reproducing. By-products from growing roots and plant residues feed soil organisms. In turn, soil organisms support plant health as they decompose organic matter, cycle nutrients, enhance soil structure and control the populations of soil organisms, both beneficial and harmful (pests and pathogens) in terms of crop productivity.

The living part of soil organic matter includes a wide variety of micro-organisms such as bacteria, viruses, fungi, protozoa and algae. It also includes plant roots, insects, earthworms, and larger animals such as moles, mice and rabbits that spend part of their life in the soil. The living portion represents about 5 per cent of the total soil organic matter. Micro-organisms, earthworms and insects help break down crop residues and manures by ingesting them and mixing them with the minerals in the soil, and in the process recycling energy and plant nutrients.

Sticky substances on the skin of earthworms and those produced by fungi and bacteria help bind particles together. Earthworm casts are also more strongly aggregated (bound together) than the surrounding soil as a result of the mixing of organic matter and soil mineral material, as well as the intestinal mucus of the worm. Thus, the living part of the soil is responsible for keeping air and water available, providing plant nutrients, breaking down pollutants and maintaining the soil structure. The composition of soil organisms depends on the food source (which in turn is season dependent). Therefore, the organisms are neither uniformly distributed through the soil nor uniformly present all year. However, in some cases their biogenic structures remain. Each species and group exists where it can find appropriate food supply, space, nutrients and moisture. Organisms occur wherever organic matter occurs. Therefore, soil organisms are concentrated: around roots, in litter, on humus, on the surface of soil aggregates and in spaces between aggregates. For this reason, they are most prevalent in forested areas and cropping systems that leave a lot of biomass on the surface.

Fig. Plate Termites Create Their Own Living Conditions Near Their Preferred Food Sources. Inside

The activity of soil organisms follows seasonal as well as daily patterns. Not all organisms are active at the same time. Most are barely active or even dormant. Availability of food is an important factor that influences the level of activity of soil organisms and thus is related to land use and management (Figure 4.8). Practices that increase numbers and activity of soil organisms include: no tillage or minimal tillage; and the maintenance of plant and annual residues that reduce disturbance of soil organisms and their habitat and provide a food supply.Different groups of organisms can be distinguished in the soil

DECOMPOSITION PROCESS

Fresh residues consist of recently deceased micro-organisms, insects and earthworms, old plant roots, crop residues, and recently added manures.Crop residues contain mainly complex carbon compounds originating from cell walls (cellulose, hemicellulose, etc.). Chains of carbon, with each carbon atom linked to other carbons, form the "backbone" of organic molecules. These carbon chains, with varying amounts of attached oxygen, H, N, P and S, are the basis for both simple sugars and amino acids and more complicated molecules of long carbon chains or rings. Depending on their chemical structure, decomposition is rapid (sugars, starches and proteins), slow (cellulose, fats, waxes and resins) or very slow (lignin).

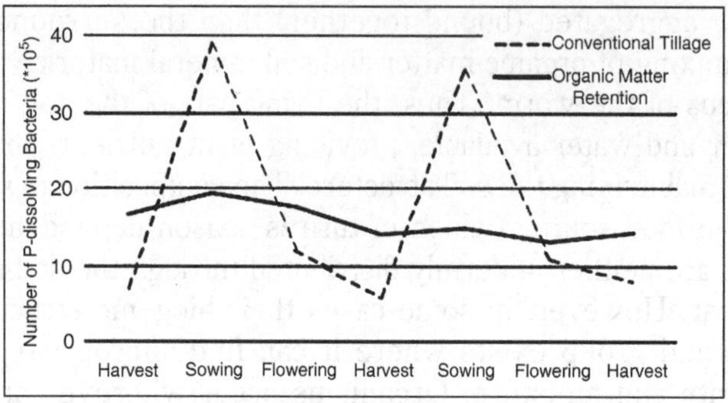

Fig. Fluctuations in Microbial Biomass at Different Stages of Crop Development in Conventional Agriculture Compared With Systems With Residue Retention and High Organic Matter Input.

Table. Classification of Soil Organisms

Micro-organisms	Microflora	<5 μm	Bacteria Fungi
	Microfauna	<100 μm	Protozoa Nematodes
Macro-organisms	Meso-organisms	100 μm-2 mm	Springtails Mites
	Macro-organisms	2-20 mm	Earthworms Millipedes Woodlice Snails and slugs
Plants	Algae Roots	10 μm > 10 μm	

Table. Essential Functions Performed by Different Members of Soil Organisms (Biota)

Functions	Organisms involved
Maintenance of soil structure	Bioturbating invertebrates and plant roots, mycorrhizae and some other micro-organisms
Regulation of soil hydrological processes	Most bioturbating invertebrates and plant roots
Gas exchange and carbon sequestration (accumulation in soil)	Mostly micro-organisms and plant roots, some C protected in large compact biogenic invertebrate aggregates
Soil detoxification	Mostly micro-organisms

Nutrient cycling	Mostly micro-organisms and plant roots, some soil- and litter-feeding invertebrates
Decomposition of organic matter	Various saprophytic and litter-feeding invertebrates (detritivores), fungi, bacteria, actinomycetes and other micro-organisms
Suppression of pests, parasites and diseases	Plants, mycorrhizae and other fungi, nematodes, bacteria and various other micro-organisms, collembola, earthworms, various predators
Sources of food and medicines	Plant roots, various insects (crickets, beetle larvae, ants, termites), earthworms, vertebrates, micro-organisms and their by-products
Symbiotic and asymbiotic relationships with plants and their roots	Rhizobia, mycorrhizae, actinomycetes, diazotrophic bacteria and various other rhizosphere micro-organisms, ants
Plant growth control (positive and negative)	Direct effects: plant roots, rhizobia, mycorrhizae, actinomycetes, pathogens, phytoparasitic nematodes, rhizophagous insects, plant-growth promoting rhizosphere micro-organisms, biocontrol agents Indirect effects: most soil biota

During the decomposition process, microorganisms convert the carbon structures of fresh residues into transformed carbon products in the soil. There are many different types of organic molecules in soil. Some are simple molecules that have been synthesized directly from plants or other living organisms. These relatively simple chemicals, such as sugars, amino acids, and cellulose are readily consumed by many organisms. For this reason, they do not remain in the soil for a long time. Other chemicals such as resins and waxes also come directly from plants, but are more difficult for soil organisms to break down.

Humus is the result of successive steps in the decomposition of organic matter. Because of the complex structure of humic substances, humus cannot be used by many micro-organisms as an energy source and remains in the soil for a relatively long time.

NON-HUMIC SUBSTANCES: SIGNIFICANCE AND FUNCTION

Non-humic organic molecules are released directly from cells of fresh residues, such as proteins, amino acids, sugars, and starches. This part of soil organic matter is the active, or easily decomposed, fraction. This active fraction is influenced strongly by weather conditions, moisture status of the soil, growth stage of the vegetation, addition of organic residues, and cultural practices, such as tillage. It is the main food supply for various organisms in the soil. Carbohydrates occur in the soil in three main forms: free sugars in the soil solution, cellulose and hemicellulose; complex polysaccharides; and polymeric molecules of various sizes and shapes that are attached strongly to clay colloids and humic substances. The simple sugars, cellulose and hemicellulose, may constitute 5-25 per cent of the organic matter in most soils, but are easily broken down by micro-organisms.

Polysaccharides (repeating units of sugar-type molecules connected in longer chains) promote better soil structure through their ability to bind inorganic soil particles into stable aggregates. Research indicates that the heavier polysaccharide molecules may be more important in promoting aggregate stability and water infiltration than the lighter molecules. Some sugars may stimulate seed germination and root elongation. Other soil properties affected by polysaccharides include CEC, anion retention and biological activity.

The soil lipids form a very diverse group of materials, of which fats, waxes and resins make up 2-6 per cent of soil organic matter. The significance of lipids arises from the ability of some compounds to act as growth hormones. Others may have a depressing effect on plant growth.

Soil N occurs mainly (> 90 per cent) in organic forms as amino acids, nucleic acids and amino sugars. Small amounts exist in the form of amines, vitamins, pesticides and their degradation products, etc. The rest is present as ammonium (NH_4^-) and is held by the clay minerals.

COMPOUNDS AND FUNCTION OF HUMUS

Humus or humified organic matter is the remaining part of organic matter that has been used and transformed by many different soil organisms. It is a relatively stable component formed by humic substances, including humic acids, fulvic acids, hymatomelanic acids and humins. It is probably the most widely distributed organic carbon-containing material in terrestrial and aquatic environments.

Humus cannot be decomposed readily because of its intimate interactions with soil mineral phases and is chemically too complex to be used by most organisms. It has many functions. One of the most striking characteristics of humic substances is their ability to interact with metal ions, oxides, hydroxides, mineral and organic compounds, including toxic pollutants, to form water-soluble and water-insoluble complexes.

Through the formation of these complexes, humic substances can dissolve, mobilize and transport metals and organics in soils and waters, or accumulate in certain soil horizons. This influences nutrient availability, especially those nutrients present at microconcentrations only. Accumulation of such complexes can contribute to a reduction of toxicity, *e.g.* of aluminium

(Al) in acid soils, or the capture of pollutants-herbicides such as Atrazine or pesticides such as Tefluthrin-in the cavities of the humic substances.

Humic and fulvic substances enhance plant growth directly through physiological and nutritional effects. Some of these substances function as natural plant hormones (auxines and gibberillins) and are capable of improving seed germination, root initiation, uptake of plant nutrients and can serve as sources of N, P and S. Indirectly, they may affect plant growth through modifications of physical, chemical and biological properties of the soil, for example, enhanced soil water holding capacity and CEC, and improved tilth and aeration through good soil structure.

About 35-55 per cent of the non-living part of organic matter is humus. It is an important buffer, reducing fluctuations in soil acidity and nutrient availability. Compared with simple organic molecules, humic substances are very complex and large, with high molecular weights. The characteristics of the well-decomposed part of the organic matter, the humus, are very different from those of simple organic molecules. While much is known about their general chemical composition, the relative significance of the various types of humic materials to plant growth is yet to be established.

Humus consists of different humic substances:

- *Fulvic acids:* The fraction of humus that is soluble in water under all pH conditions. Their colour is commonly light yellow to yellow-brown.
- *Humic acids:* The fraction of humus that is soluble in water, except for conditions more acid than pH 2. Common colours are dark brown to black.
- *Humin:* The fraction of humus that is not soluble in water at any pH and that cannot be extracted with a strong base, such as sodium hydroxide (NaOH). Commonly black in colour.

The term acid is used to describe humic materials because humus behaves like weak acids.Fulvic and humic acids are complex mixtures of large molecules. Humic acids are larger than fulvic acids. Research suggests that the different substances are differentiated from each other on the basis of their water solubility. Fulvic acids are produced in the earlier stages of humus formation. The relative amounts of humic and fulvic acids in soils vary with soil type and management practices. The humus of forest soils is characterized by a high content of fulvic acids, while the humus of agricultural and grassland areas contains more humic acids.

NATURAL FACTORS INFLUENCING THE AMOUNT OF ORGANIC MATTER

The transformation and movement of materials within soil organic matter pools is a dynamic process influenced by climate, soil type, vegetation and soil organisms. All these factors operate within a hierarchical spatial scale. Soil organisms are responsible for the decay and cycling of both macronutrients and micronutrients, and their activity affects the structure, tilth and productivity of the soil.

In natural humid and subhumid forest ecosystems without human disturbance, the living and non-living components are in dynamic equilibrium with each other. The litter on the soil surface beneath different canopy layers and high biomass production generally result in high biological activity in the soil and on the soil surface. *Mollison and Slay distinguished the following five mechanisms:*

1. A continuous soil cover of living plants, which together with the soil architecture facilitates the capture and infiltration of rainwater and protects the soil;
2. A litter layer of decomposing leaves or residues providing a continuous energy source for macro- and micro-organisms;
3. The roots of different plants distributed throughout the soil at different depths permit an effective uptake of nutrients and an active interaction with microorganisms;
4. The major period of nutrient release by micro-organisms coincides with the major period of nutrient demand by plants;
5. Nutrients recycled by deep-rooting plants and soil macrofauna and microfauna.

This equilibrium creates almost closed-cycle transfers of nutrients between soil and the vegetation adapted to such site conditions, resulting in almost perfect physical and hydric conditions for plant growth, *i.e.* a cool microclimate, increased evapotranspiration, good rooting conditions with good porosity and sufficient soil moisture. This facilitates water infiltration and prevents erosion and run-off. Thus, it results in clean water in the streams emanating from the area, a relatively smooth variation in streamflow during the year, and recharge of groundwater.

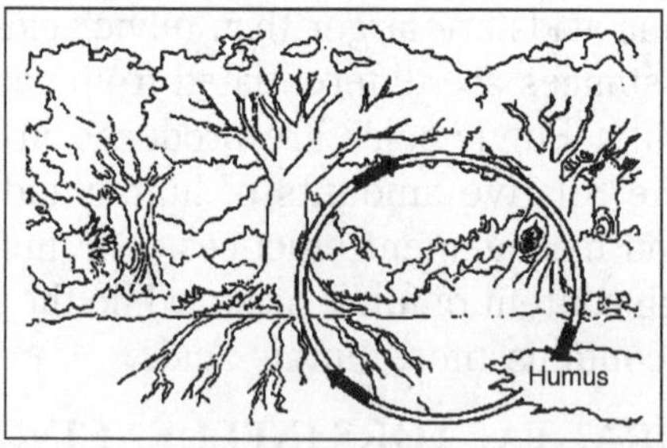

Fig. Closed Cycle

TEMPERATURE

Several field studies have shown that temperature is a key factor controlling the rate of decomposition of plant residues. Decomposition normally occurs more rapidly in the tropics than in temperate areas. Ladd and Amato reported that, despite differences in plant material and climate patterns, the decomposition of leguminous materials in southern Australian sites followed the same pattern as that of ryegrass for sites in Nigeria and the United Kingdom, although the time scales were different. Reaction rates doubled for each increase of 8-9 °C in the mean annual air temperature.

The relatively faster rate of decomposition induced by the continuous warmth in the tropics implies that high equilibrium levels of organic matter are difficult to achieve in tropical agro-ecosystems. Hence, large annual rates of organic inputs are needed to maintain an adequate labile soil organic matter pool in cultivated soils. Soils in cooler climates commonly have more organic matter because of slower mineralization (decomposition) rates.

SOIL MOISTURE AND WATER SATURATION

Soil organic matter levels commonly increase as mean annual precipitation increases. Conditions of elevated levels of soil moisture result in greater biomass production, which provides more residues, and thus more potential food for soil biota. Soil biological activity requires air and moisture.

Optimal microbial activity occurs at near "field capacity", which is equivalent to 60-per cent water-filled pore space. In the other hand, periods of water saturation lead to poor aeration. Most soil organisms need oxygen, and thus a reduction of oxygen in the soil leads to a reduction of the mineralization rate as these organisms become inactive or even die.

Some of the transformation processes become anaerobic, which can lead to damage to plant roots caused by waste products or favourable conditions for disease-causing organisms. Continued production and slow decomposition can lead to very large organic matter contents in soils with long periods of water saturation.

With the exception of the hyperhumid regions, the climates of vast areas of the humid, subhumid and semi-arid tropics are characterized by distinct wet and dry seasons. In the wet-dry tropics, large amounts of nitrate often occur in the surface soil during the first part of the rainy season. This accelerated nitrogen mineralization caused by a large increase in microbial activity is the result of the first few rains activating the labile soil organic matter.

Farmers who practise "slash and burn" agriculture often choose early planting in order to take advantage of this flush of inorganic N before it is lost through leaching and run-off. In these low-input systems, the amount of nitrate present in the soil during the early part of the rainy season is related closely to the organic matter content of the soil. N availability diminishes during the later part of the rainy season.

SOIL TEXTURE

Soil organic matter tends to increase as the clay content increases. This increase depends on two mechanisms. First, bonds between the surface of clay particles and organic matter retard the decomposition process.

Second, soils with higher clay content increase the potential for aggregate formation. Macroaggregates physically protect organic matter molecules from further mineralization caused by microbial attack.

For example, when earthworm casts and the large soil particles they contain are split by the joint action of several factors (climate, plant growth and other organisms), nutrients are released and made available to other components of soil micro-organisms. Under similar climate conditions, the organic matter content in fine textured (clayey) soils is two to four times that of coarse textured (sandy) soils.

Kaolinite, the main clay mineral in many upland soils in the tropics, has a much smaller specific surface and nutrient exchange capacity than most other clay minerals. Therefore, kaolinitic soils contain considerably fewer clay-humus complexes.

In addition, the unprotected labile humic substances are vulnerable to decomposition under appropriate soil moisture conditions. Thus, high levels of organic matter are difficult to maintain in cultivated kaolinitic soils in the wet-dry tropics, because climate and soil conditions favour rapid decomposition.

In contrast, organic matter can persist as organo-oxide complexes in soils rich in iron and aluminium oxides. Such properties favour the formation of soil microaggregates, typical of many fine-textured, oxide-rich, high base-status soils in the tropics. These soils are known for their low bulk density, high microporosity, and high organic-matter retention under natural vegetation, but also for their high phosphate fixation capacity on the oxides when used for crop production.

Current knowledge suggests that whereas organic matter contributes to the dark colour of Vertisols, it is not considered important in determining either the development, robustness or resilience of structure in these soils. Organic matter levels tend to be low in Vertisols; even as low as 10 g/ kg.

Parent material influences organic matter accumulation not only through its effect on soil texture. Soils developed from inherently rich material, such as basalt, are more fertile than soils formed from granitic material, which contains less mineral nutrients. Moreover, the former experience more organic matter accumulation because of abundant vegetative growth.

TOPOGRAPHY

Organic matter accumulation is often favoured at the bottom of hills. There are two reasons for this accumulation: conditions are wetter than at mid- or upper-slope positions, and organic matter is transported to the lowest point in the landscape through run-off and erosion. Similarly, soil organic matter levels are higher on northfacing slopes (in the Northern Hemisphere) compared with south-facing slopes (and the other way around in the Southern Hemisphere) because temperatures are lower.

SALINITY AND ACIDITY

Salinity, toxicity and extremes in soil pH (acid or alkaline) result in poor biomass production and, thus in reduced additions of organic matter to the soil. For example, pH affects humus formation in two ways: decomposition, and biomass production. In strongly acid or highly alkaline soils, the growing conditions for micro-organisms are poor, resulting in low levels of biological oxidation of organic matter.

Soil acidity also influences the availability of plant nutrients and thus regulates indirectly biomass production and the available food for soil biota. Fungi are less sensitive than bacteria to acid soil conditions.

VEGETATION AND BIOMASS PRODUCTION

The rate of soil organic matter accumulation depends largely on the quantity and quality of organic matter input. Under tropical conditions, applications of readily degradable materials with low C:N ratios, such as green manure and leguminous cover crops, favour decomposition and a short-term increase in the labile nitrogen pool during the growing season. On the other hand, applications of plant materials with both large C:N ratios and lignin contents such as cereal straw and grasses (Figure 4.11) generally favour nutrient immobilization, organic matter accumulation and humus formation, with increased potential for improved soil structure development.

Plant constituents such as lignin and other polyphenols retard decomposition. In an experiment in southern Nigeria to compare management effects on soil organic matter accumulation, a three-year fallow with Guinea grass (*Panicum maximum*), which has a high lignin content, maintained a carbon level comparable to that under forest fallow.

However, fallowing with leguminous species such as pigeon pea (*Cajanus cajan*) caused a significant decline in soil total C (Juo and Lal, 1977).

Palm and Sanchez reported that both the decomposition rate and the N-release patterns of three tropical legumes (*Inga edulis, Cajanus cajan*, and *Erythrina* spp.) were related to the amount of polyphenol compounds such as lignin in the leaf. *Erythrina* leaves had the lowest concentrations of polyphenols and the fastest decomposition rate of the three species studied.

Root turnover also constitutes an important addition of humus into the soil, and consequently it is important for carbon sequestration. In forests, most organic matter is added as superficial litter. However, in grassland ecosystems, up to two-thirds of organic matter is added through the decay of roots.

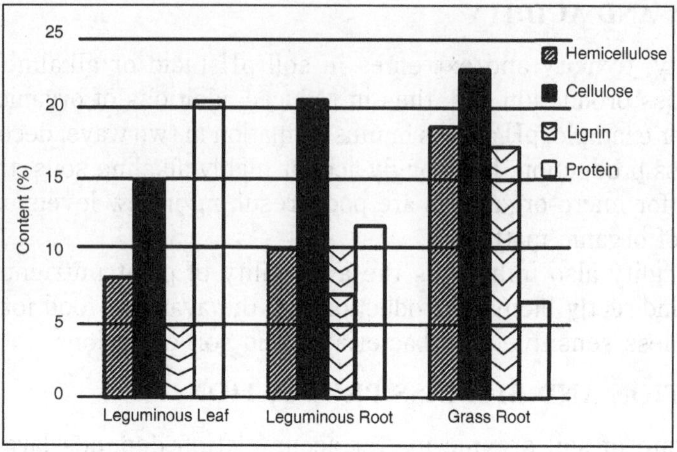

Fig. Composition of Leaves and Roots of Leguminous and Grass Species.

4

Soil and Soil Solution Chemistry

Chemical equilibrium applies to reactions that can occur in both directions. In a reaction such as:

$$CH_4(g) + H_2O(g) \leftrightarrow CO(g) + 3H_2(g)$$

The reaction can happen both ways. So after some of the products are created the products begin to react to form the reactants.

At the beginning of the reaction, the rate that the reactants are changing into the products is higher than the rate that the products are changing into the reactants. Therefore, the net change is a higher number of products.

Even though the reactants are constantly forming products and vice-versa the amount of reactants and products does become steady. When the net change of the products and reactants is zero the reaction has reached equilibrium. The equilibrium is a dynamic equilibrium. The definition for a dynamic equilibrium is when the amount of products and reactants are constant. (They are not equal but constant. Also, both reactions are still occurring.)

Some crystals of iodine, $I_2(s)$, are placed in an open gas jar (an open system) maintained at a constant room temperature. A purple vapour is seen to form above the crystals, the crystals gradually disappear, and eventually so does the purple iodine vapour. The experiment is repeated but this time a lid is placed on the gas jar (a closed system). After a short time, a uniform purple vapour forms and some crystals of iodine remain. At the constant temperature the system remains like this indefinitely.

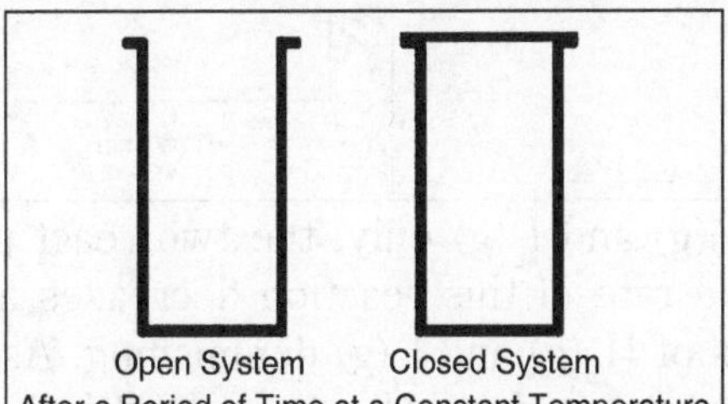

After a Period of Time at a Constant Temperature

In this closed system an equilibrium has established. The equilibrium is between the iodine solid and the iodine vapour. Here we do not have a static condition but adynamic condition. What is happening all of the time is that I_2 molecules are leaving the solid and entering the vapour phase at the same rate as they are leaving the vapour and entering the solid state. The forward and reverse rates have become the same and the system is at equilibrium. However, this is not a chemical equilibrium but a physical equilibrium.

$$I_2(s) \rightleftharpoons I_2(g)$$

Incidentally, in the closed system above, the pressure of the iodine vapour is equal to its vapour pressure. This does depend on temperature, but is independent of the volume of the vessel and of the amount of iodine solid, provided some solid is always present.

A CHEMICAL REACTION CAN ALSO REACH A STATE OF EQUILIBRIUM

At the start of an experiment you have two sealed flasks maintained at 720K, one containing a mixture of $H_2(g)$ and $I_2(g)$, and the other HI(g) only.

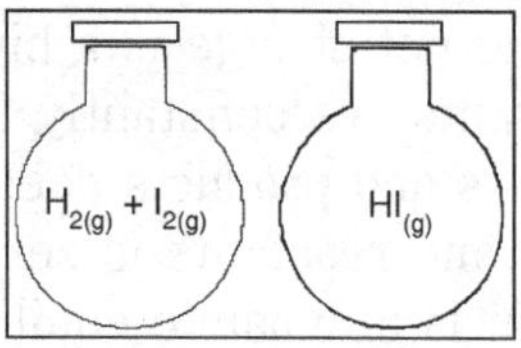

At intervals of time during the experiment the amount of HI(g) in each is measured (and expressed as the per cent of all H and I atoms present as HI).

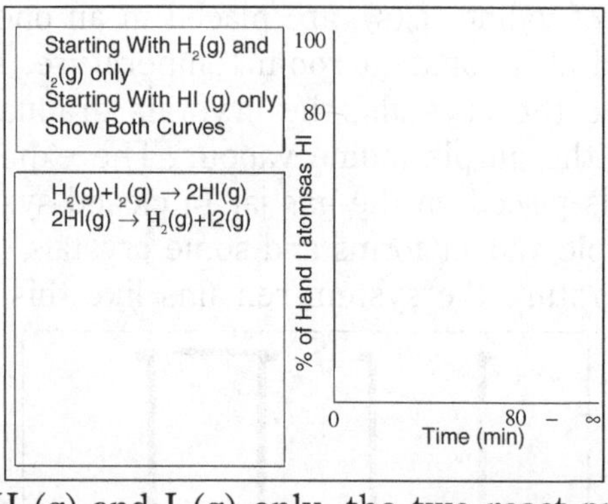

Starting with $H_2(g)$ and $I_2(g)$ only, the two react producing HI(g) in a forward reaction. The rate of this reaction decreases as it proceeds owing to the concentrations of $H_2(g)$ and $I_2(g)$ diminishing. As HI(g) is formed, its molecules begin to react to reform $H_2(g)$ and $I_2(g)$ in a reverse reaction. As the concentration of HI(g) increases as a result of the forward reaction, the rate of the reverse reaction increases. Eventually, the concentrations of all the chemical components will attain values such that the forward and reverse reactions are occuring at the same rate. We now have a reversible reaction that is said to have attained chemical equilibrium. In an equilibrium reaction, all chemical components are both reactants and products, though by tradition those shown on the left of the chemical equation are called reactants and those on the right products.

$$H_2(g) + I_2(g) \rightleftharpoons 2HI(g)$$

Remember, although at equilibrium the forward and reverse reactions may appear to have stopped they are, in fact, taking place all of the time and at the same rate-equilibrium is a dynamic not a static condition.

WHAT QUANTITIES OF REACTANTS AND PRODUCTS ARE PRESENT AT EQUILIBRIUM

Equilibrium refers to equality of forward and reverse reaction rates, not of quantities of reactants and products. In the above chemical equilibrium, 78% of the mixture is in the form of HI(g), so appreciable quantities of both reactants and products are present.

Some reactions establish equilibrium after the formation of only minute amounts of products. In others, only minute amounts of reactants remain at equilibrium. The composition of the equilibrium mixture is sometimes referred to in terms ofequilibrium position.

HOW LONG DOES IT TAKE FOR EQUILIBRIUM TO BE REACHED

In the case of HI(g), the mixture reaches equilibrium in less than 2 hours at 720 K. Other reactions may take only a tiny fraction of a second, and there are many more that would be nowhere close to equilibrium after a million years.

THE EQUILIBRIUM CONSTANT

Now consider we have a mixture of reactants and products that is not yet at equilibrium, or one that is at equilibrium. Which way will the reaction go in order to reach equilibrium? How far will the reaction go, that is, what will be the amounts of reactants and products at equilibrium?

If we make changes to the system, such as to its volume, or add more of a reactant, how will the equilibrium composition be affected? To be able to answer these questions we have to know about the Equilibrium Constant. Now back to the equilibrium involving $H_2(g)$, $I_2(g)$ and HI(g) at 720K.

$$H_2(g) + I_2(g) \rightleftharpoons 2HI(g)$$

In four separate experiments at 720 K, chemical equilibrium was allowed to establish. Two were started with different concentrations of $H_2(g)$ and

$I_2(g)$ only in each, and two with differing initial concentrations of HI(g) only. In each of the four experiments, the equilibrium concentration of $H_2(g)$, $I_2(g)$ and HI(g) were measured.

Table. Equilibrium Concentrations (Mol Dm^{-3})

$[H_2(g)]_{eqm}$	$[I_2(g)]_{eqm}$	$[HI(g)]_{eqm}$
1.14×10^{-2}	0.12×10^{-2}	2.52×10^{-2}
0.92×10^{-2}	0.20×10^{-2}	2.96×10^{-2}
0.34×10^{-2}	0.34×10^{-2}	2.35×10^{-2}
0.86×10^{-2}	0.86×10^{-2}	5.86×10^{-2}

If we substitute these equilbrium concentrations into the expression below:

$$\frac{[HI(g)]^2 \text{eqm}}{[H_2(g)]\text{eqm}[I_2(g)]\text{eqm}}$$

then a constant value, within experimental error, is obtained. The square brackets represent concentrations in mol dm^{-3}.

The value of this ratio of equilibrium concentrations (in mol dm^{-3}) is known as the equilibrium constant, represented by the symbol K_c.

$$K_c = \frac{[HI(g)]^2 \text{eqm}}{[H_2(g)]\text{eqm}[I_2(g)]\text{eqm}} (= 46.8 \text{at} 720K)$$

Many other chemical equilibrium reactions have been studied and in each case an equilibrium constant relating to the stoichiometric chemical equation has been calculated. This observation can be universally applied, in what is sometimes called the *Law of Chemical Equilibrium*, as illustrated by the general chemical equilibrium:

$$aA_{(g)} + bB_{(g)} \rightleftharpoons cC_{(g)} + dD_{(g)}$$

$$K_c = \frac{[C(g)]\overset{C}{\text{eqm}}[D(g)]\overset{d}{\text{eqm}}}{[A(g)]\overset{a}{\text{eqm}}[B(g)]\overset{b}{\text{eqm}}}$$

CAN THE EQUILBRIUM CONSTANT BE EXPRESSED IN OTHER TERMS

The Ideal Gas Equation shows that the pressure of a gas is proprtional to its concentration.

$$pV = nRT$$

where p is pressure of a particular gas (its partial pressure) in an equilibrium mixture, V is the total volume, n is the number of moles of the particular gas, R is the general gas constant, and T is the absolute temperature.

In the above equation where the temperature is also constant,

$$P \propto n/V$$

The equilibrium constant can therefore also be expressed in terms of partial pressures, and is denoted as K_p:

$$K_P = \frac{pC(g)eqm^{c} \; pD(g)eqm^{d}}{pA(g)eqm^{a} \; pB(g)eqm^{b}}$$

DOES THE EQUILIBRIUM CONSTANT HAVE UNITS

The equilibrium constant (K_c or K_p) may or may not have units; this depends precisely upon the equilibrium expression.

WHAT VALUES CAN THE EQUILIBRIUM CONSTANT HAVE

Equilibrium constants come in all sizes, from almost zero (like 10^{-50}) to extremely large (like 10^{-50}). When K (K_c or K_p) is much greater than 1, the partial pressures or concentrations of the products are large in comparison to those of the reactants.

EQUILIBRIUM CONSTANT DEPEND ON TEMPERATURE

Yes. An equilibrium constant is constant only at a constant temperature. Changing the temperature of a chemical system at equilibrium will change the value of the equilibrium constant, and also the position of the equilibrium.

ARE K_C AND K_P RELATED

Yes. The relationship depends on the reaction and how it is written. Specifically, it depends on the number of moles of *gaseous* reactants and products.

The equation is:

$$K_p = K_c(RT)^{\Delta n}$$

where Δn is the number of moles of gaseous products minus the number of moles of gaseous reactants, R is the gas constant, and T is the absolute temperature at which the equilibrium exists.

Now let's make some use of the equilibrium constant...

Return to: The Equilibrium Constant

Which way will the reaction go in order to reach equilibrium?

We have a mixture of $SO_2(g)$, $O_2(g)$, and $SO_3(g)$ at 1000K which has still to reach equilibrium. The partial pressures are p_{SO2} = 0.48 atm, p_{O2} = 0.18 atm, and p_{SO3} = 0.72 atm. K_p = 3.40 atm^{-1}. The reaction is:

$$2SO_2(g) + O_2(g) \rightleftharpoons 2SO_3(g)$$

Which way must the reaction go to reach equilibrium?

Answer:

Calculate a value (Q) for the reaction which can be compared with the equilibrium constant, K_p. This is given by the expression:

$$Q = \frac{p_{SO_3^2(g)}}{p_{SO_2^2(g)} p_{O_2(g)}}$$

The value of Q calculated is 12.50 atm^{-1}. Since Q is greater than K_p, to reach equilibrium, the equilibrium must go from right to left.

How far will the reaction go, that is, what will be the amounts of reactants and products at equilibrium?

A sample of air is compressed so that $[N_2(g)]$ = 0.80 mol dm^{-3} and $[O_2(g)]$ = 0.20 mol dm^{-3}. (It might help to imagine the container has a volume of 1 dm^3.) It is then heated to 1500 K and allowed to reach equilibrium.

$$N_2(g) + O_2(g) \rightleftharpoons 2NO(g)$$

How much of each reactant and product will be present at equilibrium? K_c = 1.0 x 10^{-5} at 1500K.

Answer:

Look at the value of Kc. Because it is such as small value, the amounts of $N_2(g)$ and $O_2(g)$ that have reacted on reaching equilibrium will be negligible compared to the original amounts present. You can assume that their equilibrium concentrations are the same as the initial concentrations. There was no NO(g) at the start, so the amount that forms will be signficant. Equal molar amounts of $N_2(g)$ and $O_2(g)$ react on reaching equilibrium, so let this amount be equal to x. From the balanced chemical equation, the amount of NO(g) at equilibrium will be 2x. Find the value of x by substituting the equilibrium concentration of each chemical component into the equilibrium expression for the reaction:

$$K_c = \frac{\left[NO(g)\right]_{eqm}^2}{\left[N_2(g)\right]_{eqm}\left[O_2(g)\right]_{eqm}} = 1.0 \times 10^{-5} \text{ at} 1500k$$

The value of x calculates to be 6.32 x 10^{-4}, so the equilibrium concentration of NO(g) is 1.26 × 10^{-3} mol dm^{-3}.

If we make changes to the system, such as to its volume, or add more of a reactant, how will the equilibrium composition be affected?

It is of great practical importance to be able to predict, and to control, the equilibrium composition of a chemical reaction.

In an industrial process it may be desirable to convert as much as possible of reactants to prodcuts. At equilibrium the product yield may not be satisfactory.

But, the composition of the equilibrium mixture (equilibrium position) is dependent on certain conditions, such as temperature and pressure, and it might be possible to change these to achieve a better yield. Le Chatelier's

Principle helps us deal with this. Now consider another chemical reaction in a state of equilibrium at a constant temperature.

$$N_2(g) + 3H_2(g) \rightleftharpoons 2NH_3(g)\ \Delta H^\circ_{f,\,298} = -\ 46.0\ kJ\ mol^{-1}$$

In a closed system and at a pressure of 250 atmospheres in the presence of a finely divided iron catalyst at 500 °C, about 15 per cent of the gases are converted into ammonia at equilibrium.

We can impose the following changes [both an increase and decrease] upon this chemical system at equilibrium:

- Amount of a substance/ concentration of a reactant or product.
- Pressure/ Volume (pressure and volume are inversely related).
- Temperature.

In general terms, this is what Le Chatelier's Principle says... If a change is imposed upon a chemical system at equilibrium then it will respond in such a way as to undo, in part, the effect of the change imposed upon it.

Here are Some Examples

Example 1

The reaction is at equilibrium in a closed vessel of fixed volume. The temperature is constant. Suddenly some nitrogen gas is added. The concentration of nitrogen has increased. The partial pressure of nitrogen is raised. The system is no longer at equilibrium. Le Chatelier's Principle says that the equilibrium will adjust so as to reduce the concentration of the nitrogen.

This can only be achieved if the nitrogen reacts. The nitrogen can only react with hydrogen forming ammonia. The equilibrium therefore shifts from left to right. A new equilibrium position is established. Since the system responds 'to undo, in part, the effect of the change', at the new equilibrium position there will be slightly more nitrogen, less hydrogen, and more ammonia present.

Example 2

The reaction is at equilibrium in a closed vessel. The temperature is constant. The volume of the vessel is suddenly reduced. This is equivalent to increasing the total pressure of the system. The system is no longer at equilibrium. Le Chatelier's Principle says that the equilibrium will adjust so as to reduce the pressure. It can only achieve this by shifting in the direction of fewer moles of gas (fewer gaseous molecules).

There are 2 moles of gas on the right of the equilibrium and 4 on the left. The equilibrium therefore shifts from left to right. A new equilibrium position is established. The yield of ammonia is increased.

Example 3

The reaction is at equilibrium in a closed vessel of fixed volume. The temperature of the system is suddenly reduced. The system is no longer at equilibrium. Le Chatelier's Principle says that the equilibrium will adjust so as to bring about a rise in temperature of the system. A reaction needs to occur so as to produce heat for this to happen. The equilibrium must shift in the exothermic direction, that is, from left to right. A new equilibrium position is established. The yield of ammonia is increased.

GETTING THE BEST YIELD OF AMMONIA

It seems that by raising the pressure of the system and at the same time reducing its temperature we can maximise the yield of ammonia at equilibrium. This is not without its problems. Maintaining a higher pressure is expensive; more robust plant (thicker pipes and stronger joints, for example) is required, and more electricity would have to be used to drive pumps to maintain this pressure. Lowering the temperature slows the rate at which chemical reactions take place resulting in a longer wait for equilibrium to be achieved. A 'compromise' has to be settled upon.

DOES A CATALYST AFFECT THE EQUILIBRIUM POSITION

No. The addition or removal of a catalyst does not cause a shift in equilibrium. A catalyst *is a substance that affects the rate of a reaction, but is not consumed in the reaction*. It does this by providing an alternative mechanism for the reaction but of lower activation enthalpy. It therefore changes the rate of approach of equilibrium and so affects the forward and reverse reaction rates in the same way. The composition of the equilibrium mixture is unchanged.

HETEROGENEOUS EQUILIBRIUM

So far we have considered only chemical equilibria involving all gases, that is, homogeneous equilibria. An equilibrium involving more than one phase-gas and solid, for example, or liquid and solid-is said to be heterogeneous. For example:

$$H_2(g) + I_2(s) \rightleftharpoons 2HI(g)$$

In the above example, the equilibrium expressions for this reaction are:

$$K_c = \frac{[HI(g)]^2 \text{ eqm}}{[H_2(g)]\text{eqm}} \quad K_p = \frac{p_{HI(g)^2 \text{eqm}}}{p_{H2(g)\text{eqm}}}$$

Notice that I_2 is missing from the equilibrium expressions. This is because, at room temperature, iodine is a solid. The composition of the equilibrium mixture of $H_2(g)$ and HI(g) is independent of the amount of solid iodine present, *as long as some of it is always present*. [The partial pressure of $I_2(g)$ (which is its vapour pressure) in equilibrium with $I_2(s)$ is constant at a constant temperature, and is independent of the volume of the container and of the quantity of solid.]

EQUILIBRIUM IN SOLUTION

Here we will consider homogeneous and heterogeneous chemical equilibria in which the solvent is much more abundant than all the other components put together. The solvent may also be a reactant or product itself.

First of all, let's look at an equilibrium that is not in solution:

$$CH_3COOH(I) + CH_3CH_2OH(I) \rightleftharpoons CH_3COOCH_2CH_3(I) + H_2O(I)$$

$$K_c = \frac{[CH_3COOCH_2CH_3(I)]eqm\ [H_2O(I)]eqm}{[CH_3CH_2OH(I)]eqm\ [CH_3COOH(I)]eqm}$$

In this case, water is not a solvent but a product only.

Now consider the same equilibrium in very dilute aqueous solution. The chemical equation is:

$$CH_3COOH(aq) + CH_3CH_2OH(aq) \rightleftharpoons CH_3COOCH_2CH_3(aq) + H_2O(l)$$

And the equilibrium expression is:

$$K_c = \frac{[CH_3COOCH_2CH_3(aq)]eqm}{[CH_3CH_2OH(aq)]eqm\ [CH_3COOH(aq)]eqm}$$

Water is now both a product and a solvent. In reaching a state of equilibrium the concentration of the water changes so little that it is effectively constant. For this reason its value is taken along with the equilibrium constant, K_c.

Here are some more examples of chemical equilibria in aqueous solution:

$$CH_3COOH_{(aq)} + H_2O_{(I)} \rightleftharpoons CH_3COO^-_{(aq)} + H_3O^+_{(aq)}$$

$$K_a = \frac{[H_3O^+(aq)]eqm[CH_3COO^-(aq)]eqm}{[CH_3COOH(aq)]eqm}$$

K_a is called the acid dissociation constant. Again, the concentration of water changes negligibly on reaching equilibrium and so its value is taken along with the equilibrium constant (K_c) to form the 'modified' equilibrium constant, K_a.

$$NH_{3(aq)} + H_2O_{(l)} \rightleftharpoons NH_4^+{}_{(aq)} + OH^-_{(aq)}$$

$$K_b = \frac{[NH_4^+(aq)]eqm[OH^-(aq)]eqm}{[NH_3(aq)]eqm}$$

K_b is called the base dissociation constant. Again, the concentration of water changes negligibly on reaching equilibrium and so its value is taken along with the equilibrium constant (K_c) to form the 'modified' equilibrium constant, K_b.

$$AgCl_{(s)} \rightleftharpoons Ag^{+}_{(aq)} + Cl^{-}_{(aq)}$$

$$K_s = \left[Ag^{+}_{(aq)}\right]eqm\left[Cl^{-}_{(aq)}\right]eqm$$

K_s is the Solubility Product. It is the equilibrium constant for the equilibrium that exists between a slightly soluble salt and its ions in saturated solution. The solid, AgCl, is omitted from the equilibrium expression. As long as there is some solid present the concentration of the saturated solution of ions remains constant at a given temperature.

Chemical change is one of the two central concepts of chemical science, the other being *structure*. The very origins of Chemistry itself are rooted in the observations of transformations such as the combustion of wood, the freezing of water, and the winning of metals from their ores that have always been a part of human experience.

It was, after all, the quest for some kind of constancy underlying change that led the Greek thinkers of around 200 BCE to the idea of elements and later to that of the atom.

It would take almost 2000 years for the scientific study of matter to pick up these concepts and incorporate them into what would emerge, in the latter part of the 19th century, as a modern view of chemical change.

CHEMICAL CHANGE: HOW FAR, HOW FAST

Chemical change occurs when the atoms that make up one or more substances rearrange themselves in such a way that new substances are formed. These substances are the *components* of the *chemical reaction system*; those components which decrease in quantity are called *reactants*, while those that increase are *products*.

A given chemical reaction system is defined by a *balanced net chemical equation* which is conventionally written as *reactants* → *products* The first thing we need to know about a chemical reaction represented by a balanced equation is whether it can actually take place. If the reactants and products are all substances capable of an independent existence, then in principle, the answer is always "yes". This answer must be qualified, however, by the following considerations:

That is, what fraction of the reactants are converted into products? Some reactions convert essentially 100% of reactants to products, while for others the quantity of products may be undetectable. Many are somewhere in between, meaning that significant quantities of all components remain at the end. Later on, in another part of the course, you will learn that the tendency of a reaction to occur can be predicted entirely from the properties of the reactants and products through the laws of thermodynamics.

Some reactions are over in microseconds; others take years. The speed of any one reaction can vary over a huge range depending on the temperature, the state of matter (gas,liquid, solid) and the presence of a catalyst.

Unlike the question of completeness, there is no simple way of predicting reaction speed. What happens, at the atomic or molecular level, when reactants are transformed into products? What intermediate species (those that are produced but later consumed so that they do not appear in the net reaction equation) are involved? This is the *microscopic*, or kinetic view of chemical change, and cannot be predicted by theory as it is presently developed and must be inferred from the results of experiments.

A reaction that is thermodynamically possible but for which no reasonably rapid mechanism is available is said to be ***kinetically limited***. Conversely, one that occurs rapidly but only to a small extent is ***thermodynamically limited***. As you will see later, there are often ways of getting around both kinds of limitations, and their discovery and practical applications constitute an important area of industrial chemistry.

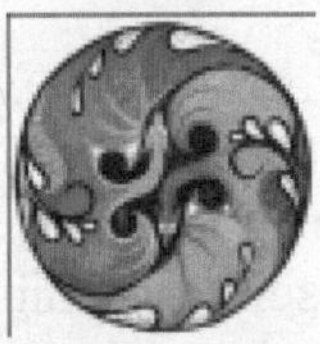

Fig. Textile Artist Hilary Rice Offers This View of Equilibrium Which Nicely Evokes its Dual Character of Internal Dynamics Contained Within a Static Exterior.

Basically, the term refers to what we might call a "balance of forces". In the case of mechanical equilibrium, this is its literal definition.

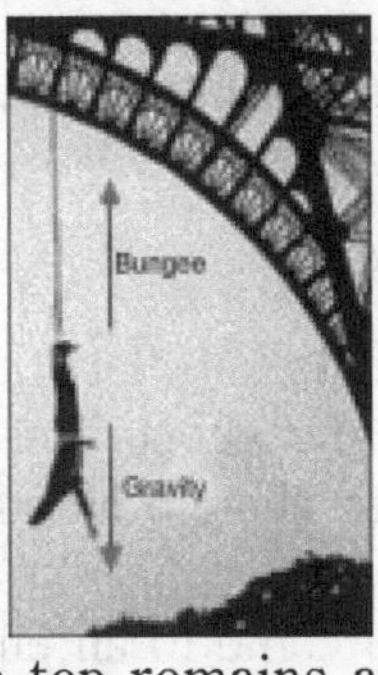

A book sitting on a table top remains at rest because the downward force exerted by the earth's gravity acting on the book's mass (this is what is meant by the "weight" of the book) is exactly balanced by the repulsive force between atoms that prevents two objects from simultaneously occupying the same space, acting in this case between the table surface.

An object is in a state of mechanical equilibrium when it is either static (motionless) or in a state of unchanging motion.

From the relation $f = ma$, it is apparent that if the net force on the object is zero, its acceleration must also be zero, so if we can see that an object is not undergoing a change in its motion, we know that it is in mechanical equilibrium.

THERMAL EQUILIBRIUM

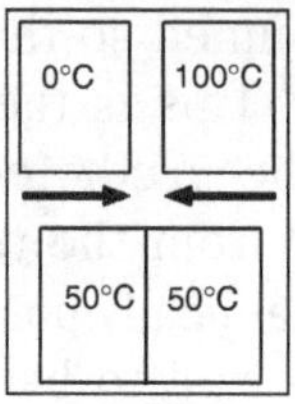

Another kind of equilibrium we all experience is thermal equilibrium. When two objects are brought into contact, heat will flow from the warmer object to the cooler one until their temperatures become identical. Thermal equilibrium arises from the tendency of thermal energy to become as dispersed or "diluted" as possible.

A metallic object at room temperature will feel cool to your hand when you first pick it up because the thermal sensors in your skin detect a flow of heat from your hand into the metal, but as the metal approaches the temperature of your hand, this sensation diminishes.

The time it takes to achieve thermal equilibrium depends on how readily heat is conducted within and between the objects; thus a wooden object will feel warmer than a metallic object even if both are at room temperature because wood is a relatively poor thermal conductor and will therefore remove heat from your hand more slowly.

Thermal equilibrium is something we often want to avoid, or at least postpone; this is why we insulate buildings, perspire in the summer and wear heavier clothing in the winter.

CHEMICAL EQUILIBRIUM

When a chemical reaction takes place in a container which prevents the entry or escape of any of the substances involved in the reaction, the quantities of these components change as some are consumed and others are formed.

Eventually this change will come to an end, after which the composition will remain unchanged as long as the system remains undisturbed. The system is then said to be in its *equilibrium state*, or more simply, "at equilibrium".

Why reactions go towards equilibrium

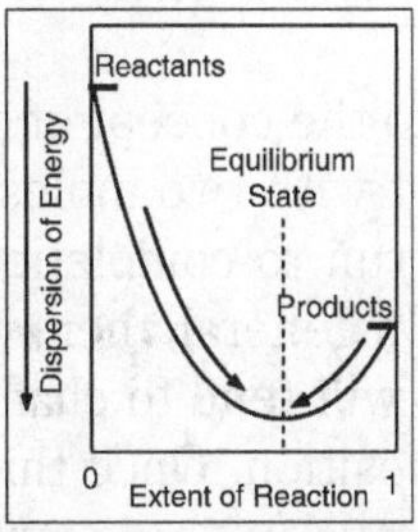

What is the nature of the "balance of forces" that drives a reaction towards chemical equilibrium? It is essentially the balance struck between the tendency of energy to reside within the chemical bonds of stable molecules, and its tendency to become dispersed and diluted. Exothermic reactions are particularly effective in this, because the heat released gets dispersed in the infinitely wider world of the surroundings. In the reaction represented here, this balance point occurs when about 60% of the reactants have been converted to products. Once this equilibrium state has been reached, no further net change will occur. (The only spontaneous changes that are allowed follow the arrows pointing towards maximum dispersal of energy.) A more complete explanation of this must be deferred until the discussion of thermodynamics in a later chapter.

Fig. Chemical Equilibrium is Something You Definitely Want to Avoid For Yourself as Long as Possible.

The myriad chemical reactions in living organisms are constantly moving *toward*equilibrium, but are prevented from getting there by input of reactants and removal of products. So rather than being in equilibrium, we try to maintain a "steady-state" condition which physiologists call *homeostasis*—maintenance of a constant internal environment.

For the time being, it's very important that you know this definition: A chemical reaction is in equilibrium when there is no tendency for the quantities of reactants and products to change. The direction in which we write a chemical reaction (and thus which components are considered reactants and which are products) is arbitrary.

Thus the two equations:

1. $H_2 + I_2 \rightarrow 2HI$ "Synthesis of hydrogen iodide"
2. $2\ HI \rightarrow H_2 + I_2$ "Dissociation of hydrogen iodide"

represent the same chemical reaction system in which the roles of the components are reversed, *and both yield the same mixture of components when the change is completed.*

This last point is central to the concept of chemical equilibrium. It makes no difference whether we start with two moles of HI or one mole each of H_2 and I_2; once the reaction has run to completion, the quantities of these two components will be the same. In general, then, we can say that thecomposition of a chemical reaction system will tend to change in a direction that brings it closer to its equilibrium composition. Once this equilibrium composition has been attained, no further change in the quantities of the components will occur as long as the system remains undisturbed.

The figure 8.10 show how the concentrations of the three components of this chemical reaction change with time. Examine the two sets of plots carefully, noting which substances have zero initial concentrations, and are thus "products" of the reaction equations shown. Satisfy yourself that these two sets represent the same *chemical reaction system*, but with the reactions occurring in opposite directions. Most importantly, note how the final (equilibrium) concentrations of the components are the same in the two cases.

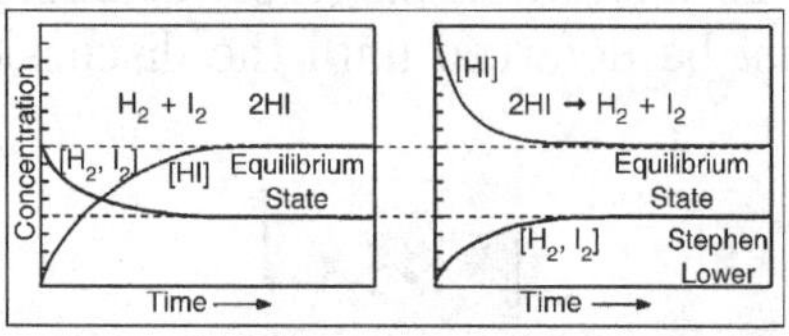

Whether we start with an equimolar mixture of H_2 and I_2 (left) or a pure sample of hydrogen iodide (shown on the right, using twice the initial concentration of HI to keep the number of atoms the same), the composition after equilibrium is attained (shaded regions on the right) will be the same.

The equilibrium composition is independent of the direction from which it is approached.

What is a Reversible Reaction?

A chemical equation of the form $A \rightarrow B$ represents the transformation of A into B, but it does not imply that *all* of the reactants will be converted into products, or that the *reverse* reaction $B \rightarrow A$ cannot also occur.

In general, *both* processes (forward and reverse) can be expected to occur, resulting in an *equilibrium mixture* containing finite amounts of *all* of the components of the reaction system. (We use the word *components* when we do not wish to distinguish between reactants and products.)

If the equilibrium state is one in which significant quantities of both reactants and products are present, then the reaction is said to *incomplete* or *reversible*. The latter term is preferable because it avoids confusion with "complete" in its other sense of being completed or finished, implying that the reaction has run its course and is now at equilibrium.

- If it is desired to emphasize the reversibility of a reaction, the single arrow in the equation is replaced with a pair of hooked lines pointing in opposite directions, as in $A \rightleftharpoons B$.
  - Note that there is no fundamental difference between the meanings of $A \rightarrow B$ and $A \rightleftharpoons B$. Some older textbooks just use A = B.
- A reaction is said to be *complete* or *quantitative* when the equilibrium composition contains no significant amount of the reactants. However, a reaction that is complete when written in one direction is said "not to occur" when written in the reverse direction.

In principle, all chemical reactions are reversible, but this reversibility may not be observable if the fraction of products in the equilibrium mixture is very small, or if the reverse reaction is very slow (the chemist's term is "*kinetically inhibited*") How did Napoleon Bonaparte help discover reversible reactions?

Napoleon recruited the eminent French chemist Claude Louis Berthollet (1748-1822) to accompany him as scientific advisor on the most far-flung of his campaigns, the expedition into Egypt in 1798. Once in Egypt, Berthollet noticed deposits of sodium carbonate around the edges of some the salt lakes found there. He was already familiar with the reaction,

$$Na_2CO_3 + CaCl_2 \rightarrow CaCO_3 + 2NaCl$$

He immediately realised that the Na_2CO_3 must have been formed by the reverse of this process brought about by the very high concentration of salt in the slowly-evaporating waters. This led Berthollet to question the belief of the time that a reaction could only proceed in a single direction. His famous textbook *Essai de statique chimique* (1803) presented his speculations on chemical affinity and his discovery that an excess of the product of a reaction could drive it in the reverse direction. Unfortunately, Berthollet got a bit carried away by the idea that a reaction could be influenced by the amounts of substances present, and maintained that the same should be true for the compositions of individual compounds. This brought him into conflict with the recently accepted Law of Definite Proportions (that a compound is made up of fixed numbers of its constituent atoms), so his ideas (the good along with the bad) were promptly discredited and remained largely forgotten for 50 years. (Ironically, it is now known that certain classes of compounds do in fact exhibit variable composition of the kind that Berthollet envisioned.)

What is the Law of Mass Action?

Berthollet's ideas about reversible reactions were finally vindicated by experiments carried out by others, most notably the Norwegian chemists (and brothers-in-law) Cato Guldberg and Peter Waage. During the period 1864-1879 they showed that an equilibrium can be approached from either direction, implying that any reaction $aA + bB \rightarrow cC + dD$ is really a competition between a "forward" and a "reverse" reaction. When a reaction is at equilibrium, the rates of these two reactions are identical, so no *net* (macroscopic) change is observed, although individual components are actively being transformed at the microscopic level.

Equilibrium is Dynamic

The "active masses" are essentially the*concentrations* of the reactants and products that combine directly in the manner represented by the reaction equation. The meaning of the terms on the right sides of the equations is simply that the rate is proportional to the concentrations of the components.

Guldberg and Waage showed that for a reaction $aA + bB \rightarrow cC + dD$, the rate (speed) of the reaction in either direction is proportional to what they called the "active masses" of the various components:

$$\text{Rate of forward reaction} = k_f[A]^a[B]^b$$
$$\text{Rate of reverse reaction} = k_r[C]^c[D]^d$$

in which the proportionality constants *k* are called *rate constants* and the quantities in square brackets represent concentrations. If we combine the two reactants A and B, the forward reaction starts immediately; then, as the products C and D begin to build up, the reverse process gets underway. As the reaction proceeds, the rate of the forward reaction diminishes while that of the reverse reaction increases. Eventually the two processes are proceeding at the same rate, and the reaction is at equilibrium:

$$\text{Rate of forward reaction} = \text{rate of reverse reaction}$$
$$k_f[A]^a[B]^b = k_r[C]^c[D]^d$$

It is very important that you understand the significance of this relation. The equilibrium state is one in which there is no net change in the quantities of reactants and products. But do not confuse this with a state of "no change"; at equilibrium, the forward and reverse reactions continue, but *at identical rates*, essentially cancelling each other out. Equilibrium is macroscopically static, but is microscopically dynamic!

To further illustrate the dynamic character of chemical equilibrium, suppose that we now change the composition of the system previously at equilibrium by adding some C or withdrawing some A (thus changing their "active masses"). The reverse rate will temporarily exceed the forward rate and a change in composition ("a shift in the equilibrium") will occur until a new equilibrium composition is achieved. The composition of the equilibrium state depends on the ratio of the forward- and reverse rate constants. Be sure you understand the difference between the *rate* of a reaction and a *rate constant*. The latter, usually designated by k, relates the reaction rate to the concentration of one or more of the reaction components—for example, *rate* = k [A]. At equilibrium the rates of the forward and reverse processes are identical, but the rate *constants* are generally different. To see how this works, consider the simplified reaction A → B in the following three scenarios.

$$k_f \gg k_r$$

$$\underbrace{K_f \times [A]}_{\text{rate of forward reaction}} = \underbrace{K_r \times [B]}_{\text{rate of reverse reaction}}$$

If the rate constants are greatly different (by many orders of magnitude), then this requires that the equilibrium concentrations of products exceed those of the reactants by the same ratio. Thus the equilibrium composition will lie strongly on the "right"; the reaction can be said to be "complete" or "quantitative".

$$k_f \ll k_r$$

$$\underbrace{K_f \times [A]}_{\text{rate of forward reaction}} = \underbrace{K_r \times [B]}_{\text{rate of reverse reaction}}$$

The rates can only be identical if the concentrations of the products are very small. We describe the resulting equilibrium as strongly favoring the left; very little product is formed. In the most extreme cases, we might even say that "the reaction does not take place".

$$k_f \approx k_r$$

If k_f and k_r have comparable values then signficant concentrations of products and reactants are present at equilibrium; we say the the reaction is "incomplete" and "reversible". The images shown in figure 8.12 offer yet another way of looking at these three cases. The plots show how the relative concentrations of the reactant and product change during the course of the reaction. The plots differ in the assumptions we make about the ratio of k_f to k_r. The equilibrium composition of the system is illustrated by the proportions of A and B in the horizontal parts of each plot where the composition remains unchanged. In each case, the two rate constants are sufficiently close in magnitude that each reaction can be considered "incomplete".

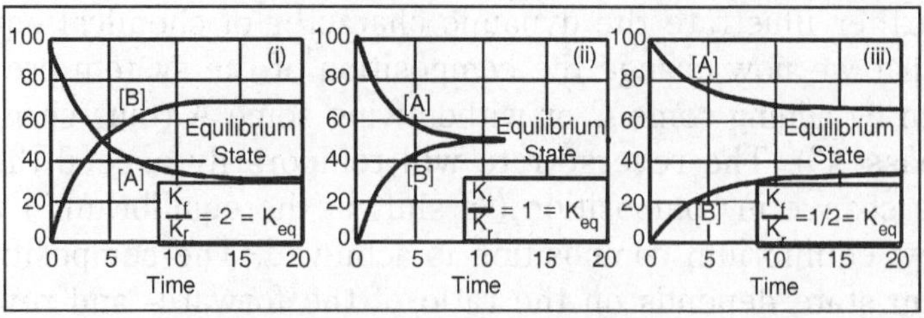

Fig. Here For a Very Nice Online Simulation That Allows You to Explore The Effects of Your own Set of Starting Concentrations and Rate Constants on This Kind of Plot.

1. In plot *(i)* the forward rate constant is twice as large as the reverse rate constant, so product (B) is favoured, but there is sufficient reverse reaction to maintain a significant quantity of A.
2. In *(ii)*, the forward and reverse rate constants have identical magnitudes. Not surprisingly, so are the equilibrium values of [A] and [B].
3. In *(iii)*, the reverse rate constant exceeds the forward rate constant, so the eqilibrium composition is definitely "on the left".

The Law of Mass Action is thus essentially the statement that the equilibrium composition of a reaction mixture can vary according to the quantities of components that are present. This of course is just what Berthollet observed in his Egyptian salt ponds, but we now understand it to be a consequence of the dynamic nature of chemical equilibrium.

How do we know when a Reaction is at Equilibrium?

Clearly, if we observe some change taking place— a change in colour, the release of gas bubbles, the appearance of a precipitate, or the release of heat, we know the reaction is not yet at equilibrium. But the absence of any apparent change does not by itself establish that the reaction is at equilibrium. The equilibrium state is one in which not only no change in composition take place, but also one in which no energetic tendency for further change is present. Unfortunately, "tendency" is not a property that is directly observable! Consider, for example, the reaction representing the synthesis of water from its elements:

$$2\ H_2(g) + O_2(g) \rightarrow 2\ H_2O(g)$$

You can store the two gaseous reactants in the same container indefinitely without any observable change occurring. But if you create an electrical spark in the container or introduce a flame, *bang*! After you pick yourself up off the floor and remove the shrapnel from what's left of your body, you will know very well that the system was not initially at equilibrium! It happens that this particular reaction has a tremendous tendency to take place, but for reasons that we will discuss in a later chapter, nothing can happen until we "set it off" in some way— in this case by exposing the mixture to a flame or spark, or (in a more gentle way) by introducing a platinum wire, which acts as a *catalyst*. A reaction of this kind is said to be highly favoured thermodynamically, but inhibited kinetically. The similar reaction of hydrogen and iodine

$$H_2(g) + I_2(g) \rightarrow 2\ HI(g)$$

by contrast is only moderately favoured thermodynamically (and is thus incomplete), but its kinetics are both unspectacular and reasonably facile.

Some Simple Tests for the Equilibrium State

- As we explained above in the context of the law of mass action, addition or removal of one component of the reaction will affect the amounts of all the others. For example, if we add more of a reactant, we would expect to see the concentration of a product change. If this does not happen, then it is likely that the reaction is kinetically inhibited and that the system is unable to attain equilibrium.
- It is almost always the case, however, that once a reaction actually starts, it will continue on its own until it reaches equilibrium, so if we can observe the change as it occurs and see it slow down and stop, we can be reasonably certain that the system is in equilibrium. This is by far the chemist's most common criterion.
- There is one other experimental test for equilibrium in a chemical reaction, although it is really only applicable to the kind of reactions we described above as being reversible. As we shall see later, the equilibrium state of a system is always sensitive to the temperature, and often to the pressure, so any changes in these variables, however, small, will temporarily disrupt the equilibrium, resulting in an observable change in the composition of the system as it moves towards its new equilibrium state.

SUMMARY

It is especially imortant that you know the precise meanings of all the highlighted terms in the context of this topic:

- Any reaction that can be represented by a balanced chemical equation can take place, at least in principle. However, there are two important qualifications:
  1. The tendency for the change to occur may be so small that the quantity of products formed may be very low, and perhaps negligible. A reaction of this kind is said to be *thermodynamically inhibited*. The tendency for chemical change is governed solely by the properties of the reactants and products, and can be predicted by applying the laws of thermodynamics.

2. The rate at which the reaction proceeds may be very small, or even zero, in which case we say the reaction is *kinetically inhibited*. Reaction rates depend on the *mechanism* of the reaction— that is, on what actually happens to the atoms as reactants are transformed into products. Reaction mechanisms cannot generally be predicted, and must be worked out experimentally. Also, the same reaction may have different mechansims under different conditions.

- As a chemical change proceeds, the quantities of the components on one side of the reaction equation will decrease, and those on the other side will increase. Eventually the reaction slows down and the composition of the system stops changing. At this point the reaction is in its *equilibrium state*, and no further change in composition will occur as long as the system is left undisturbed.
- For many reactions, the equilibrium state is one in which components on both sides of the equation (that is, both reactants and products) are present in significant amounts. Such a reaction is said to be *incomplete* or *reversible*.
- The equilibrium composition is independent of the direction from which it is approached; the labeling of substances as "reactants" or "products" is entirely a matter of convenience.
- The *law of mass action* states that any chemical change is a competition between a *forward* reaction (left-to-right in the chemical equation) and a *reverse* reaction. The rate of each of these processes is governed by the concentrations of the substances reacting; as the reaction proceeds, these rates approach each other and at equilibrium they become identical.
- From the above, it follows that equilibrium is a *dynamic process* in which *microscopic change* (the forward and reverse reactions) continues to occur, but*macroscopic change* (changes in the quantities of substances) is absent.
- When a chemical reaction is at equilibrium, any disturbance of the system, such as a change in temperature, or addition or removal of one of the reaction components, will "shift" the composition to a new equilibrium state. This is the only unambiguous way of verifying that a reaction is at equilibrium. The fact that the composition remains static does not in itself prove that a reaction is at equilibrium, because the change may be kinetically inhibited.

EQUILIBRIUM CONSTANT

To determine the amount of each compound that will be present at equilibrium you must know the equilibrium constant. To determine the equilibrium constant you must consider the generic equation:

$$aA + bB \leftrightarrow cC + dD$$

The upper case letters are the molar concentrations of the reactants and products. The lower case letters are the coefficients that balance the equation. Use the following equation to determine the equilibrium constant (K_c).

$$K_c = \frac{[C]^c [D]^d}{[A]^a [B]^b}$$

For example, determining the equilibrium constant of the following equation can be accomplished by using the K_c equation. Using the following equation, calculate the equilibrium constant.

$$N_2(g) + 3H_2(g) \leftrightarrow 2NH_3(g)$$

A one-liter vessel contains 1.60 moles NH_3.800 moles N_2, and 1.20 moles of H_2. What is the equilibrium constant?

$$K_c = \frac{[1.60]^2}{[.800][1.20]^3} = 29.6$$

Answer: 1.85

For a general *elementary* chemical reaction,

$$aA + bB \leftrightarrow cC + dD$$

The concentrations of the reactants and products are related to each other according to,

$$K_c = \frac{[C]^c [D]^d}{[A]^a [B]^b}$$

The number K_c is called the ***equilibrium constant***, and is a function of temperature only (*i.e.*, its numerical value doesn't change unless the temperature changes–we'll hold the temperature constant for now). Note the word *elementary*, more on that later; for now, all reactions are elementary. The stoichiometric coefficients a, b, c and d show up as powers of the corresponding reactants (and products–just call them all reactants from now on). For our example reaction,

$$A + B \leftrightarrow C$$

the equilibrium constant is defined,

$$K_c = \frac{[C]}{[A][B]}$$

Notice that the units of K_c in this case are M^{-1}= L/mol.

There is another useful definition of the equilibrium constant based on pressure rather than concentration. The ideal gas law reads,

$$PV = nRT$$

Here P is the total pressure. In the case of several components, each has a partial pressure, all of which sum up to the total pressure:

$$P = P_A + P_a + P_c$$

For each component, we can write the ideal gas law (putting a subscript where applicable)

$$P_A V = n_A RT \Rightarrow \frac{n_A}{V} = [A] = \frac{P_A}{RT}$$

This works for the other components, too, and gives us the relation between the concentrations and the partial pressures. If we plug in the partial pressures in the definition of K_c above, we get,

$$K_c = \frac{[C]}{[A][B]} = \frac{P_c}{RT}\frac{RT}{P_A}\frac{RT}{P_B} = \frac{P_c}{P_A P_B}(RT) = K_P(RT)$$

In the more general case where the reaction is,

$$aA + bB \leftrightarrow cC + dD$$

And K_c is,

$$K_c = \frac{[C]^c[D]^d}{[A]^a[B]^b}$$

Then the new equilibrium constant becomes,

$$K_P = K_c(RT)^{(c+d)-(a+b)} = K_c(RT)^{\Delta n}$$

Where,

$$\Delta n = (c+d) - (a+b)$$

Is the change in number of moles in the gas phase. These numbers come from the balanced equation and are basically the sum of the stoichiometric coefficients of the products minus the sum of the stoichiometric coefficients of the reactants.

Problems using K_P instead of K_c are solved the same way, except we are looking for the partial pressures P_X instead of the concentrations [X], X = A, B, C, etc. is particularly useful for gas phase problems since the pressure is usually more convenient to measure than the concentration. Also, there is a direct relationship between K_P and the Gibbs energy of the reaction, $\Delta G^0{}_{rxn}$.

LE CHATELIER'S PRINCIPLE

Le Chatelier's principle states that when a system in chemical equilibrium is disturbed by a change of temperature, pressure, or a concentration, the system shifts in equilibrium composition in a way that tends to counteract this change of variable.

The three ways that Le Chatelier's principle says you can affect the outcome of the equilibrium are as follows:

1. Changing concentrations by adding or removing products or reactants to the reaction vessel.
2. Changing partial pressure of gaseous reactants and products.
3. Changing the temperature.

These actions change each equilibrium differently, therefore you must determine what needs to happen for the reaction to get back in equilibrium.

Example involving change of concentration:

In the equation,

$$2NO_{(g)} + O_{2(g)} \leftrightarrow 2NO_{2(g)}$$

- If you add more $NO_{(g)}$ the equilibrium shifts to the right producing more $NO_{2(g)}$
- If you add more $O_{2(g)}$ the equilibrium shifts to the right producing more $NO_{2(g)}$
- If you add more $NO_{2(g)}$ the equilibrium shifts to the left producing more $NO_{(g)}$ and $O_{2(g)}$

Example involving pressure change:

In the equation,

$$2SO_{2(g)} + O_{2(g)} \leftrightarrow 2SO_{3(g)},$$

an increase in pressure will cause the reaction to shift in the direction that reduces pressure, that is the side with the fewer number of gas molecules. Therefore an increase in pressure will cause a shift to the right, producing more product. (A decrease in volume is one way of increasing pressure.)

Example involving temperature change:

In the equation,

$$N_{2(g)} + 3H_{2(g)} \leftrightarrow 2NH_3 + 91.8 \text{ kJ},$$

an increase in temperature will cause a shift to the left because the reverse reaction uses the excess heat. An increase in forward reaction would produce even more heat since the forward reaction is exothermic. Therefore the shift caused by a change in temperature depends upon whether the reaction is exothermic or endothermic. If a reaction is at equilibrium and we alter the conditions so as to create a new*equilibrium state*, then the composition of the system will tend to change until that new equilibrium state is attained. (We say "tend to change" because if the reaction is ***kinetically inhibited***, the change may be too slow to observe or it may never take place.)

In 1884, the French chemical engineer and teacherHenri Le Châtelier (1850–1936) showed that in every such case, the new equilibrium state is one that partially reduces the effect of the change that brought it about.This law is known to every Chemistry student as the *Le Châtelier principle*. His original formulation was somewhat complicated, but a reasonably useful paraphrase of it reads as follows: To see how this works (and you *must* do so, as this is of such fundamental importance that you simply cannot do any meaningful chemistry without a thorough working understanding of this principle), look again at the hydrogen iodide dissociation reaction

$$2HI \rightarrow H_2 + I_2$$

Consider an arbitrary mixture of these three components at equilibrium, and assume that we inject more hydrogen gas into the container. Because the H_2 concentration now exceeds its new equilibrium value, the system is no longer in its equilibrium state, so a net reaction now ensues as the system moves to the new state.

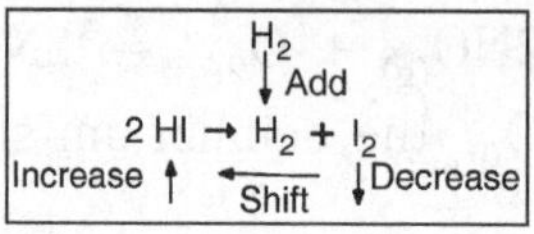

The Le Châtelier principle states that the net reaction will be in a direction that tends to reduce the effect of the added H_2. This can occur if some of the H_2 is consumed by reacting with I_2 to form more HI; in other words, a net reaction occurs in the reverse direction. Chemists usually simply say that "the equilibrium shifts to the left".

To get a better idea of how this works, carefully examine the diagram below which follows the concentrations of the three components of this reaction as they might change in time (the time scale here will typically be about an hour):

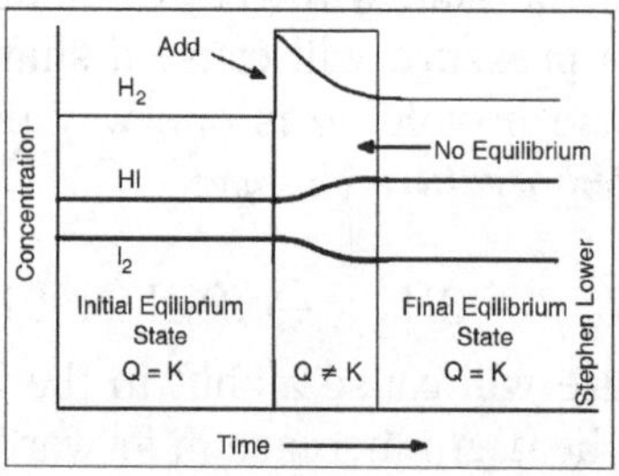

DISRUPTION AND RESTORATION OF EQUILIBRIUM

At the left, the concentrations of the three components do not change with time because the system is at equilibrium. We then add more hydrogen to the system, disrupting the equilibrium. A net reaction then ensues that moves the system to a new equilibrium state (right) in which the quantity of hydrogen iodide has increased; in the process, some of the I_2 and H_2 are consumed.

Notice that the new equilibrium state contains more hydrogen than did the initial state, but not as much as was added; as the le Châtelier principle predicts, the change we made (addition of H_2) has been partially counteracted by the "shift to the right".

The following tabLe Contains several examples showing how changing the quantity of a reaction component can shift an established equilibrium.

System	Change	Result
$CO_2 + H_2 \rightarrow H_2O(g) + CO$	a drying agent is added to absorb H_2O	**Shift to the right**. Continuous removal of a product will force any reaction to the right
$H_2(g) + I_2(g) \rightarrow 2HI(g)$	Some nitrogen gas is added	**No change**; N_2 is not a component of this reaction system.
$NaCl(s) + H_2SO_4(l) \rightarrow Na_2SO_4(s) + HCl(g)$	reaction is carried out in an open container	Because HCl is a gas that can escape from the system, the reaction is forced to the **right**. This is the basis for the commercial production of hydrochloric acid.
$H_2O(l) \rightarrow H_2O(g)$	water evaporates from an open container	Continuous removal of water vapor forces the reaction to the **right**, so equilibrium is never achieved.
$HCN(aq) \rightarrow H^+(aq) + CN^-(aq)$	the solution is diluted	Shift to **right**; the product $[H^+][CN^-]$ diminishes more rapidly than does [HCN].
$AgCl(s) \rightarrow Ag^+(aq) + Cl^-(aq)$	some NaCl is added to the solution	Shift to **left** due to increase in Cl^- concentration. This is known as the common ion effect on solubility.
$N_2 + 3 H_2 \rightarrow 2 NH_3$	a catalyst is added to speed up this reaction	**No change**. Catalysts affect only the rate of a reaction; the have no effect at all on the composition of the equilibrium state.

HOW DO CHANGES IN TEMPERATURE AFFECT EQUILIBRIA

Virtually all chemical reactions are accompanied by the liberation or uptake of heat. If we regard heat as a "reactant" or "product" in an endothermic or exothermic reaction respectively, we can use the Le Chatelier principle to predict the direction in which an increase or decrease in temperature will shift the equilibrium state. Thus for the oxidation of nitrogen, an endothermic process, we can write,

$$[\text{heat}] + N_2 + O_2 \rightarrow 2\,NO$$

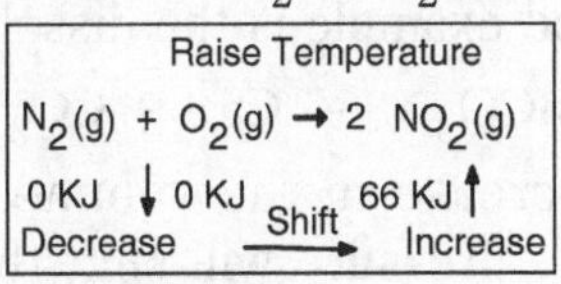

Suppose this reaction is at equilibrium at some temperature T_1 and we raise the temperature to T_2. The Le Châtelier principle tells us that a net reaction will occur in the direction that will partially counteract this change. Since the reaction is endothermic, a shift of the equilibrium to the right will take place. Nitric oxide, the product of this reaction, is a major air pollutant which initiates a sequence of steps leading to the formation of atmospheric smog. Its formation is an unwanted side reaction which occurs when the air (which is introduced into the combustion chamber of an engine to supply oxygen) gets heated to a high temperature. Designers of internal combustion engines now try, by various means, to limit the temperature in the combustion region, or to restrict its highest-temperature part to a small volume within the combustion chamber.

HOW DO CHANGES IN PRESSURE AFFECT EQUILIBRIA

You will recall that if the pressure of a gas is reduced, its volume will increase; pressure and volume are inversely proportional. With this in mind, suppose that the reaction

$$2\ NO_2(g) \rightarrow N_2O_4(g)$$

is in equilibrium at some arbitrary temperature and pressure, and that we double the pressure, perhaps by compressing the mixture to a smaller volume. From the Le Châtelier principle we know that the equilibrium state will change to one that tends to counteract the increase in pressure. This can occur if some of the NO_2 reacts to form more of the dinitrogen tetroxide, since two moles of gas are being removed from the system for every mole of N_2O_4 formed, thereby decreasing the total volume of the system. Thus increasing the pressure will shift this equilibrium to the right.

It is important to understand that changing the pressure will have a significant effect only on reactions in which there is a change in the number of moles of gas.

For the above reaction, this change

$$\Delta n_g = (n_{products} - n_{reactants}) = 1\text{–}2 = -1.$$

In the case of the nitrogen oxidation reaction,

$N_2 + O_2 \rightarrow 2\ NO$, $\Delta n_g = 0$ and pressure will have no effect.

The volumes of solids and liquids are hardly affected by the pressure at all, so for reactions that do not involve gaseous substances, the effects of pressure changes are ordinarily negligible. Exceptions arise under conditions of very high pressure such as exist in the interior of the Earth or near the bottom of the ocean. A good example is the dissolution of calcium carbonate,

$$CaCO_3(s) \rightarrow Ca^{2+} + CO_3^{2-}.$$

There is a slight decrease in the volume when this reaction takes place, so an increase in the pressure will shift the equilibrium to the right, with the results that calcium carbonate becomes more soluble at higher pressures.

The skeletons of several varieties of microscopic organisms that inhabit the top of the ocean are made of $CaCO_3$, so there is a continual rain of this substance towards the bottom of the ocean as these organisms die. As a consequence, the floor of the Atlantic ocean is covered with a blanket of calcium carbonate.

This is not true for the Pacific ocean, which is deeper; once the skeletons fall below a certain depth, the higher pressure causes them to dissolve. Some of the seamounts (undersea mountains) in the Pacific extend above the solubility boundary so that their upper parts are covered with $CaCO_3$ sediments.

The effect of pressure on a reaction involving substances whose boiling points fall within the range of commonly encountered temperature will be sensitive to the states of these substances at the temperature of interest. For reactions involving gases, only changes in the partial pressures of those gases directly involved in the reaction are important; the presence of other gases has no effect.

Problem Example

The commercial production of hydrogen is carried out by treating natural gas with steam at high temperatures and in the presence of a catalyst ("steam reforming of methane"):

$$CH_4 + H_2O \rightarrow CH_3OH + H_2$$

Given the following boiling points: CH_4 (methane) = –161°C, H_2O = 100°C, CH_3OH = 65°, H_2 = –253°C, predict the effects of an increase in the total pressure on this equilibrium at 50°, 75° and 120°C.

Solution: Calculate the change in the moles of gas for each process:

Temp	Equation	D n_g	Shift
50°	$CH_4(g) + H_2O(l)$ ® $CH_3OH(l) + H_2(g)$	0	none
75°	$CH_4(g) + H_2O(l)$ ® $CH_3OH(g) + H_2(g)$	+1	to left
120°	$CH_4(g) + H_2O(g)$ ® $CH_3OH(g) + H_2(g)$	0	none

WHAT IS THE HABER PROCESS AND WHY IS IT IMPORTANT

The Haber process for the synthesis of ammonia is based on the exothermic reaction:

$$N_2(g) + 3\,H_2(g) \rightarrow 2\,NH_3(g) \quad \Delta H = -92 \text{ kJ/mol}$$

The Le Chatelier principle tells us that in order to maximize the amount of product in the reaction mixture, it should be carried out at high pressure and low temperature. However, the lower the temperature, the slower the reaction (this is true of virtually all chemical reactions.) As long as the choice had to be made between a low yield of ammonia quickly or a high yield over a long period of time, this reaction was infeasible economically. Nitrogen is available for free, being the major component of air, but the strong triple bond in

N_2 makes it extremely difficult to incorporate this element into species such as NO_3^- and NH_4^+ which serve as the starting points for the wide variety of nitrogen-containing compounds that are essential for modern industry.

This conversion is known as *nitrogen fixation*, and because nitrogen is an essential plant nutrient, modern intensive agriculture is utterly dependent on huge amounts of fixed nitrogen in the form of fertilizer. Until around 1900, the major source of fixed nitrogen was the $NaNO_3$ found in extensive deposits in South America. Several chemical processes for obtaining nitrogen compounds were developed in the early 1900's, but they proved too inefficient to meet the increasing demand.

$$CH_4 + H_2O \rightarrow CO + 3H_2 \text{ formation of synthesis gas from methane}$$

$$CO + H_2O \rightarrow CO_2 + H_2 \textit{ shift reaction} \text{ carried out in reformer}$$

The Haber-Bosch process is considered the most important chemical synthesis developed in the 20th century. Besides its scientific importance as the first large-scale application of the laws of chemical equilibrium, it has had tremendous economic and social impact; without an inexpensive source of fixed nitrogen, the intensive crop production required to feed the world's growing population would have been impossible. Haber was awarded the 1918 Nobel Prize in Chemistry in recognition of his

THE LE CHATELIER PRINCIPLE IN PHYSIOLOGY

Many of the chemical reactions that occur in living organisms are regulated through the Le Châtelier principle.

OXYGEN TRANSPORT BY THE BLOOD

Few of these are more important to warm-blooded organisms than those that relate to aerobic respiration, in which oxygen is transported to the cells where it is combined with glucose and metabolized to carbon dioxide, which then moves back to the lungs from which it is expelled.

$$\text{Hemoglobin} + O_2 \rightleftharpoons \text{Oxyhemoglobin}$$

The partial pressure of O_2 in the air is 0.2 atm, sufficient to allow these molecules to be taken up by hemoglobin (the red pigment of blood) in which it becomes loosely bound in a complex known as oxyhemoglobin. At the ends of the capillaries which deliver the blood to the tissues, the O_2 concentration is reduced by about 50% owing to its consumption by the cells.

This shifts the equilibrium to the left, releasing the oxygen so it can diffuse into the cells.

Much more detail about the mechanism by which hemoglobin transfers oxygen from the lungs to tissues, and then carries carbon dioxide and hydrogen ions back to the lungs, can be found on this U. of Virginia page.

MAINTENCE OF BLOOD PH

Carbon dioxide reacts with water to form a weak acid H_2CO_3 which would cause the blood pH to fall to dangerous levels if it were not promptly removed as it is excreted by the cells. This is accomplished by combining it with carbonate ion through the reaction,

$$H_2CO_3 + CO_3^{2-} \rightleftharpoons 2HCO_3^-$$

which is forced to the right by the high local CO_2 concentration within the tissues. Once the hydrogen carbonate (bicarbonate) ions reach the lung tissues where the CO_2 partial pressure is much smaller, the reaction reverses and the CO_2 is expelled.

CARBON MONOXIDE POISONING

Carbon monoxide, a product of incomplete combustion that is present in automotive exhaust and cigarette smoke, binds to hemoglobin 200 times more tightly than does O_2. This blocks the uptake and transport of oxygen by setting up a competing equilibrium

$$O_2\text{-hemoglobin} \rightleftharpoons \text{hemoglobin} \rightleftharpoons \text{CO-hemoglobin}$$

Air that contains as little as 0.1 per cent carbon monoxide can tie up about half of the hemoglobin binding sites, reducing the amount of O_2 reaching the tissues to fatal levels. Carbon monoxide poisoning is treated by administration of pure O_2 which promotes the shift of the above equilibrium to the left. This can be made even more effective by placing the victim in a hyperbaric chamber in which the pressure of O_2 can be made greater than 1 atm.

Make sure you thoroughly understand the following essential ideas which have been presented above. It is especially imortant that you know the precise meanings of all the highlighted terms in the context of this topic.

- A system in its *equilibrium state* will remain in that state indefinitely as long as it is undisturbed. If the equilibrium is destroyed by subjecting the system to a change of pressure, temperature, or the number of moles of a substance, then a net reaction will tend to take place that moves the system to a new equilibrium state. *Le Châtelier's principle* says that this net reaction will occur in a direction that partially offsets the change.
- The Le Châtelier Principle has practical effect only for reactions in which signficant quantities of both reactants and products are present at equilibrium— that is, for reactions that are *thermodynamically reversible*.
- Addition of more product substances to an equilibrium mixture will shift the equilibrium to the left; addition of more reactant substances

will shift it to the right. These effects are easily explained in terms of competing forward- and reverse reactions—that is, by the *law of mass action*.

- If a reaction is *exothermic* (releases heat), an increase in the temperature will force the equilibrium to the left, causing the system to absorb heat and thus partially ofsetting the rise in temperature. The opposite effect occurs for*endothermic* reactions, which are shifted to the right by rising temperature.
- The effect of pressure on an equilibrium is significant only for reactions which involve different numbers of moles of gases on the two sides of the equation. If the number of moles of gases increases, than an increase in the total pressure will tend to initiate a reverse reaction that consumes some the products, partially reducing the effect of the pressure increase.
- The classic example of the practical use of the Le Châtelier principle is the *Haber-Bosch process* for the synthesis of ammonia, in which a balance between low temperature and high pressure must be found.

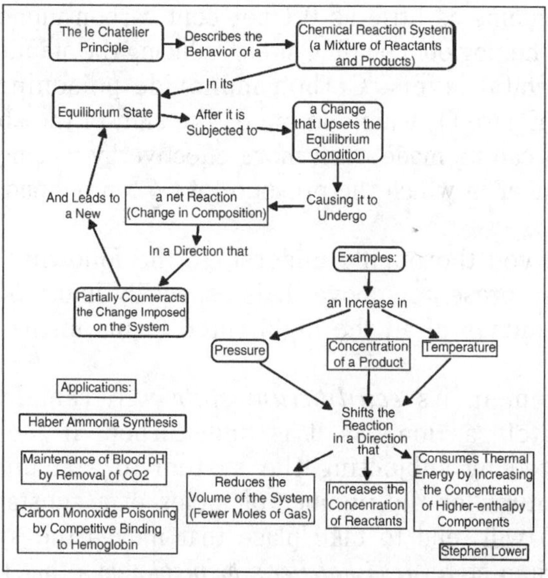

Q AND K: WHAT'S THE DIFFERENCE

Given a chemical reaction system and its equilibrium constant, together with the concentrations or pressures of the reactants and products actually present, we often need to predict whether, and in which direction, a net change in composition will tend to take place. The concept of the reaction quotient makes this very easy to do, but the similarity of the algebraic forms of K and Q can easily lead to confusion.

WHAT IS THE EQUILIBRIUM QUOTIENT

We defined the *equilibrium expression* for the reaction

$$a\,A + b\,B \rightarrow c\,C + d\,D \text{ as}$$

$$\frac{[C]^c\,[D]^d}{[A]^a\,[B]^b}$$

In the general case in which the concentrations can have any arbitrary values (including zero), this expression is called the *reaction quotient*(the term *equilibrium quotient* is also commonly used.) and its value is denoted by Q (or Q_c or Q_p if we wish to emphasize that the terms represent molar concentrations or partial pressures.) If the terms correspond to *equilibrium* concentrations, then the above expression is called the *equilibrium constant* and its value is denoted by K (or K_c orK_p.)K is thus the special value that Q has when the reaction is at equilibrium

The value of Q in relation to K serves as an index how the composition of the reaction system compares to that of the equilibrium state, and thus it indicates the direction in which any net reaction must proceed.For example, if we combine the two reactants A and B at concentrations of 1 mol L^{-1} each, the value of Q will be $0 \div 1 = 0$. The only possible change is the conversion of some of these reacants into products. If instead our mixture consists only of the two products C and D, Q will be indeterminately large $(1 \div 0)$ and the only possible change will be in the reverse direction.

It is easy to see (by simple application of the le Châtelier principle) that the ratio of Q/K immediately tells us whether, and in which direction, a net reaction will occur as the system moves towards its equilibrium state. A schematic view of this relationship is shown below:

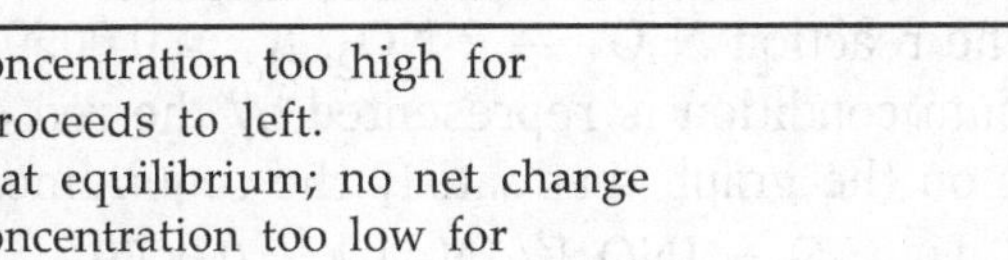

More formally, we simply look at which of the following describes the relative values of Q and K:

Q/K		
> 1	Product concentration too high for reaction proceeds to left.	equilibrium; net
= 1	System is at equilibrium; no net change	will occur.
< 1	Product concentration too low for reaction proceeds to	equilibrium; net right.

It is very important that you be able to work out these relations for yourself, not by memorizing them, but from the definitions of Q and K.

Problem Example

The equilibrium constant for the oxidation of sulfur dioxide is $K_p = 0.14$ at 900 K.

$$2\ SO_2(g) + O_2(g) \rightarrow 2\ SO_3(g)$$

If a reaction vessel is filled with SO_3 at a partial pressure of 0.10 atm and with O_2 and SO_2 each at a partial pressure of 0.20 atm, what can you conclude about whether, and in which direction, any net change in composition will take place?

Solution:

The value of the equilibrium quotient Q for the initial conditions is,

$$Q = \frac{(PSO_3)^2}{(PO_2)(PSO_2)^2} = \frac{(0.10\,\text{atm})^2}{(0.20\,\text{atm})(0.20\,\text{atm})^2} = 1.25\,\text{atm}^{-1}$$

Since $Q > K$, the reaction is not at equilibrium, so a net change will occur in a direction that decreases Q. This can only occur if some of the SO_3 is converted back into products. In other words, the reaction will "shift to the left".

A VISUAL WAY OF THINKING ABOUT *Q* AND *K*

The formal definitions of Q and K are quite simple, but they are of limited usefulness unless you are able to relate them to real chemical situations.

The following diagrams illustrate the relation between Q and K from various standpoints. Take some time to study each one carefully, making sure that you are able to relate the description to the illustration.

Example: Dissociation of Dinitrogen Tetroxide

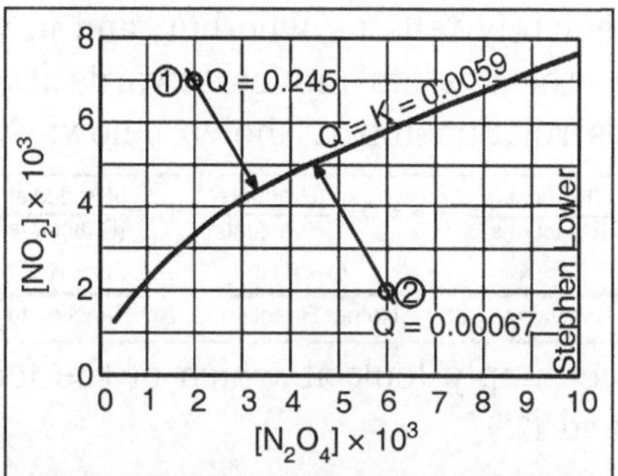

For the reaction $N_2O_4 \rightarrow 2\ NO_2$, $K_c = 0.0059$ at 298K.

This equilibrium condition is represented by the red curve that passes through all points on the graph that satisfy the requirement that,

$$Q = [NO_2]^2 / [N_2O_4] = 0.0059.$$

There are of course an infinite number of possible Q's of this system within the concentration boundaries shown on the plot. Only those points that fall on the red line correspond to *equilibrium states* of this system (those for which $Q = K$). The line itself is a plot of $[NO_2]$ that we obtain by rearranging the equilibrium expression $[NO_2] = ([N_2O_4]K)^{0.5}$.

If the system is initially in a non-equilibrium state, its composition will tend to change in a direction that moves it to one that is on the line.

Two such non-equilibrium states are shown. The state indicated by has 1. $Q > K$, so we would expect a net reaction that reduces Q by converting some of the NO_2 into N_2O_4; in other words, the equilibrium "shifts to the left". Similarly, in state 2. $Q < K$, indicating that the forward reaction will occur.

The arrows in the figure 8.16 indicate the successive values that Q assumes as the reaction moves closer to equilibrium. The slope of the line reflects the stoichiometry of the equation. In this case, one mole of reactant yields two moles of products, so the slopes have an absolute value of 2:1.

Example 2: Phase-Change Equilibrium

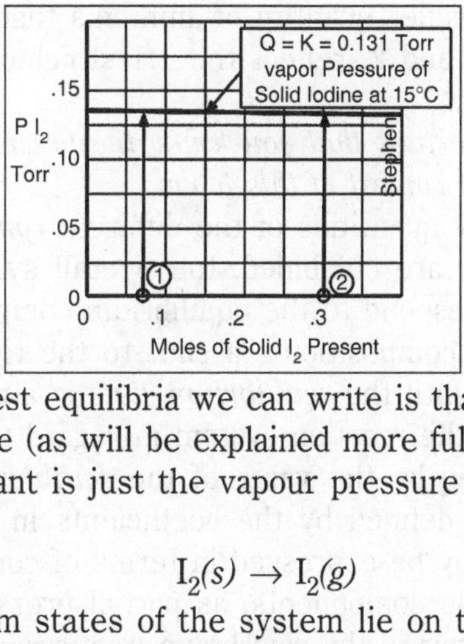

One of the simplest equilibria we can write is that between a solid and its vapour. In this case (as will be explained more fully in the next lesson), the equilibrium constant is just the vapour pressure of the solid. Thus for the process,

$$I_2(s) \rightarrow I_2(g)$$

all possible equilibrium states of the system lie on the horizontal red line and is independent of the quantity of solid present (as long as there is at least enough to supply the relative tiny quantity of vapour.)

So adding various amounts of the solid to an empty closed vessel (states 1. and 2.) causes a gradual buidup of iodine vapour. Because the equilibrium pressure of the vapour is so small, the amount of solid consumed in the process is negligible, so the arrows go straight up and all lead to the same equilibrium vapour pressure.

Example: Heterogeneous Chemical Reaction

The decomposition of ammonium chloride is a common example of a heterogeneous (two–phase) equilibrium. Solid ammonium chloride has a substantial vapour pressure even at room temperature:

$$NH_4Cl(s) \rightarrow NH_3(g) + HCl(g)$$

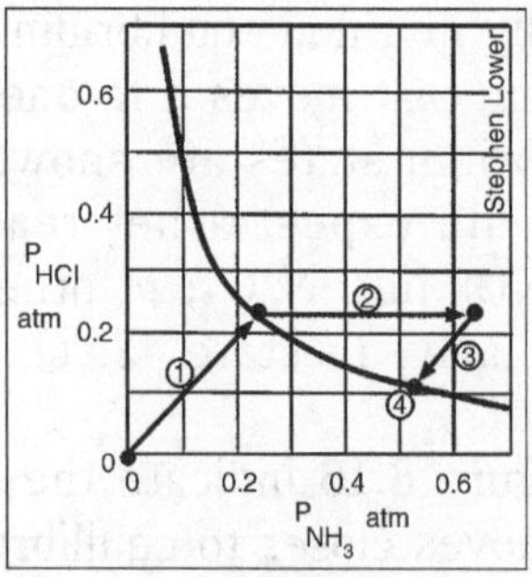

Arrow 1. traces the states the system passes through when solid NH_4Cl is placed in a closed container. Arrow 2. represents the addition of ammonia to the equilibrium mixture; the system responds by following the path 3. back to a new equilibrium state 4. which, as the Le Châtelier principle predicts, contains a smaller quantity of ammonia than was added. The unit slopes of the paths 1. and 2. reflect the 1:1 stoichiometry of the gaseous products of the rection.

It is especially imortant that you know the precise meanings of all the highlighted terms in the context of this topic:

- When arbitrary quantities of the different *components* of a *chemical reaction system* are combined, the overall system composition will not likely correspond to the equilibrium composition. As a result, a net change in composition ("a shift to the right or left") will tend to take place until the *equilibrium state* is attained.
- The status of the reaction system in regard to its equilibrium state is characterised by the value of the *equilibrium expression* whose formulation is defined by the coefficients in the balanced reaction equation; it may be expressed in terms of concentrations, or in the case of gaseous components, as partial pressures.
- The various terms in the equilibrium expression can have any arbitrary value (including zero); the value of the equilibrium expression itself is called the*reaction quotient Q*.
- If the concentration or pressure terms in the equilibrium expression correspond to the equilibrium state of the system, then Q has the special value K, which we call the *equilibrium constant*.
- The ratio of Q/K (whether it is 1, >1 or <1) thus serves as an index of how far the system is from its equilibrium composition, and its value indicates the direction in which the net reaction must proceed in order to reach its equilibrium state.
- When $Q = K$, then the equilibrium state has been reached, and no further net change in composition will take place as long as the system remains undisturbed.

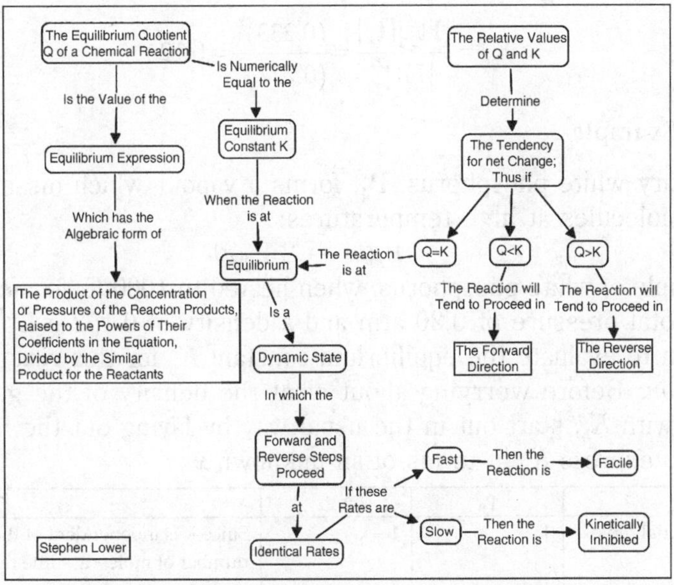

EQUILIBRIUM CALCULATIONS

MEASURING AND CALCULATING EQUILIBRIUM CONSTANTS

Clearly, if the concentrations or pressures of all the components of a reaction are known, then the value of K can be found by simple substitution. Observing individual concentrations or partial pressures directly may be not always be practical, however. If one of the components is coloured, the extent to which it absorbs light of an appropriate wavelength may serve as an index of its concentration. Pressure measurements are ordinarily able to measure only the total pressure of a gaseous mixture, so if two or more gaseous products are present in the equilibrium mixture, the partial pressure of one may need to be inferred from that of the other, taking into account the stoichiometry of the reaction.

Problem Example

In an experiment carried out by Taylor and Krist (*J. Am. Chem. Soc.* 1941: 1377), hydrogen iodide was found to be 22.3% dissociated at 730.8°K. Calculate K_c for $2\ HI(g) \rightarrow H_2(g) + I_2$.

Solution:

No explicit molar concentrations are given, but we do know that for every *n*moles of HI, 0.223*n* moles of each product is formed and (1–0.233)*n* = 0.777*n* moles of HI remains. For simplicity, we assume that $n = 1$ and that the reaction is carried out in a 1.00-L vessel, so that we can substitute the required concentration terms directly into the equilibrium expression.

$$K_c = \frac{[H_2][I_2]}{[HI]^2} = \frac{(0.233)^2}{(0.777)^2} = 0.12$$

Problem Example

Ordinary white phosphorus, P_4, forms a vapour which dissociates into diatomic molecules at high temperatures:

$$P_4(g) \rightarrow 2\,P_2(g)$$

A sample of white phosphorus, when heated to 1000°C, formed a vapour having a total pressure of 0.20 atm and a density of 0.152 g L^{-1}. Use this information to evaluate the equilibrium constant K_p for this reaction.

Solution: Before worrying about what the density of the gas mixture has to do with K_p, start out in the usual way by laying out the information required to express K_p in terms of an unknown x.

	P_4	$2\,P_2$	
Initial moles:	1	1 – x	Since *K* is independent of the number of moles, assume the simplest case.
Moles at equilibrium:	1 – x	2x	*x* is the fraction of P_4 that dissociates.
Eq. mole fractions:	$\left(\frac{1-x}{1+x}\right)$	$\left(\frac{2x}{1+x}\right)$	The denominator is the total number of moles: $(1-x) + 2x = 1+x$.
Eq. partial pressures:	$\left(\frac{1-x}{1+x}\right)\times 0.2$	$\left(\frac{2x}{1+x}\right)\times 0.2$	Partial pressure is the mole fraction times the total pressure.

Expressing the equilibrium constant in terms of x gives,

$$K_P = \frac{P_{P2}^2}{P_{P4}} = \frac{\left(\frac{1-x}{1+x}\right)^2}{\left(\frac{2x}{1+x}\right)} = \frac{(1-x)^2}{2x}$$

Now we need to find the dissociation fraction x of P_4, and at this point we hope you remember those gas laws that you were told you would be needing later in the course! The density of a gas is directly proportional to its molecular weight, so you need to calculate the densities of pure P_4 and pure P_2 vapors under the conditions of the experiment. One of these densities will be greater than 0.152 gL^{-1} and the other will be smaller; all you need to do is to find where the measured density falls in between the two limits, and you will have the dissociation fraction. The molecular weight of phosphorus is 31.97, giving a molar mass of 127.9 g for P_4. This mass must be divided by the volume to find the density; assuming ideal gas behaviour, the volume of 127.9 g (1 mole) of P_4 is given by RT/P, which works out to 522 L (remember to use the absolute temperature here.) The density of pure P_4 vapour under the conditions of the experiment is then

$d = m/V = (128\ g\ mol^{-1}) \times = (522\ L\ mol^{-1}) = 0.245\ g\ L^{-1}$.

The density of pure P_2 would be half this, or 0.122 g L^{-1}. The difference between these two limiting densities is 0.123 g L^{-1}, and the difference between the density of pure P_4 and that of the equilibrium mixture is (.245 –.152) g L^{-1}or 0.093 g L^{-1}. The ratio 0.093 0.123 = 0.76 is therefore the fraction of P_4that remains and its fractional dissociation is (1–0.76) = 0.24. Substituting into the equilibrium expression gives K_p = 1.2.

PREDICTING EQUILIBRIUM CONCENTRATIONS

This is by far the most common kind of equilibrium problem you will encounter: starting with an arbitrary number of moles of each component, how many moles of each will be present when the system comes to equilibrium?

The principal source of confusion and error for beginners relates to the need to determine the values of several unknowns (a concentration or pressure for each component) from a single equation, the equilibrium expression.

The key to this is to make use of the stoichiometric relationships between the various components, which usually allow us to express the equilibrium composition in terms of a single variable.

The easiest and most error-free way of doing this is adopt a systematic approach in which you create and fill in a small table as shown in the following problem example. You then substitute the equilibrium values into the equilibrium constant expression, and solve it for the unknown.

This very often involves solving a quadratic or higher-order equation. Quadratics can of course be solved by using the familiar quadratic formula, but it is often easier to use an algebraic or graphical approximation, and for higher-order equations this is the only practical approach.

There is almost never any need to get an exact answer, since the equilibrium constants you start with are rarely known all that precisely anyway.

Problem Example

Phosgene ($COCl_2$) is a poisonous gas that dissociates at high temperature into two other poisonous gases, carbon monoxide and chlorine. The equilibrium constant K_p = 0.0041 at 600°K. Find the equilibrium composition of the system after 0.124 atm of $COCl_2$ is allowed to reach equilibrium at this temperature.

Solution: Start by drawing up a table showing the relationships between the components:

	$COCl_2$	CO	Cl_{2}®	
Initial pressures:	0.124 atm		0	0
Change:	–x	+x	+x	
Equilibrium pressures:	0.124–x	x	x	

Substitution of the equilibrium pressures into the equilibrium expression gives,

$$\frac{x^2}{0.124-x}=0.0041$$

This expression can be rearranged into standard polynomial form $x^2 +.0041\,x–0.00054 = 0$ and solved by the quadratic formula, but we will simply obtain an approximate solution by iteration. Because the equilibrium constant is small, we know that x will be rather small compared to 0.124, so the above relation can be approximated by,

$$\frac{x^2}{0.124}=0.0041$$

which gives $x = 0.0225$. To see how good this is, substitute this value of x into the denominator of the original equation and solve again:

$$\frac{x^2}{0.124-0.0225}=\frac{x^2}{0.102}=0.0041$$

This time, solving for x gives 0.0204. Iterating once more, we get,

$$\frac{x^2}{0.124-0.0204}=\frac{x^2}{0.104}=0.0041$$

and $x = 0.0206$ which is sufficiently close to the previous to be considered the final result. The final partial pressures are then 0.104 atm for $COCl_2$, and 0.0206 atm each for CO and Cl_2.

Problem Example

The gas-phase dissociation of phosphorus pentachloride to the trichloride has Kp = 3.60 at 540°C:

$$PCl_5 \rightarrow PCl_3 + Cl_2$$

What will be the partial pressures of all three components if 0.200 mole of $PCl_5$and 3.00 moles of PCl_3 are combined and brought to equilibrium at this temperature and at a total pressure of 1.00 atm?

	PCl_5	PCl_3	Cl_2
Initial moles:	0.200	3.00	0
Change:	–x	+x	+x
Equilibrium moles:	0.200 – x	3.00 + x	x
Eq. partial pressures:	$\left(\frac{2.00-x}{3.20+x}\right)$	$\left(\frac{3.00+x}{3.20+x}\right)$	$\left(\frac{x}{3.20+x}\right)$

Solution: As always, set up a table showing what you know (first two rows) and then expressing the equilibrium quantities:

The partial pressures in the bottom row were found by multiplying the mole fraction of each gas by the total pressure: $P_i = X_i P_t$.

The term in the denominator of each mole fraction is the total number of moles of gas present at equilibrium: $(0.200–x) + (3.00 + x) + x = 3.20 + x$. Substituting the equilibrium partial pressures into the equilibrium expression, we have,

$$\frac{(3.00+x)(x)}{(0.200-x)(3.20+x)} = 3.60$$

whose polynomial form is $4.60\,x^2 + 13.80\,x–2.304 = 0$.

Plotting this on a graphical calculator yields x = 0.159 as the positive root:

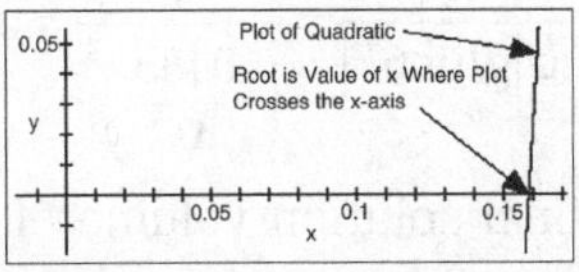

Substitution of this root into the expressions for the equilibrium partial pressures in the table yields the following values: $P(PCl_5) = 0.012$ atm, $P(PCl_3) = 0.94$ atm, $P(Cl_2) = 0.047$ atm.

EFFECTS OF DILUTION ON EQUILIBRIUM

In this the LeChâtelier principle, it was mentioned that diluting a weak acid such as acetic acid CH_3COOH ("HAc") will shift the dissociation equilibrium to the right:

$$HAc + H_2O \rightarrow H_3O^+ + Ac^-$$

Thus a 0.10*M* solution of acetic acid is 1.3% ionized, while in a 0.01M solution, 4.3% of the HAc molecules will be dissociated. This is because as the solution becomes more dilute, the product $[H_3O^+][Ac^-]$ decreases more rapidly than does the [HAc] term. At the same time the concentration of H_2O becomes greater, but because it is so large to start with (about 55.5M), any effect this might have is negligible, which is why no $[H_2O]$ term appears in the equilibrium expression. For a reaction such as $CH_3COOH(I) + C_2H_5OH(I) \rightarrow CH_3COOC_2H_5(I) + H_2O(I)$ (in which the water concentration does change), dilution will have no effect on the equilibrium; the situation is analogous to the way the pressure dependence of a gas-phase reaction depends on the number of moles of gaseous components on either side of the equation.

Problem Example

The biochemical formation of a disaccharide (double) sugar from two monosaccharides is exemplified by the reaction, fructose + glucose-6-phosphate → sucrose-6-phosphate

(Sucrose is ordinary table sugar.) To what volume should a solution containing 0.050 mol of each monosaccharide be diluted in order to bring about 5% conversion to sucrose phosphate?

Solution: The initial and final numbers of moles are as follows: td>

	Fructose	**Glucose–6–P**	**Sucrose–6–P**
Initial moles:	0.05	0.05	0
Equilibrium moles:	0.0485	0.0485	0.0015

Substituting into the expression for K_c in which the solution volume is the unknown, we have

$$\frac{|\text{suc6P}|}{|\text{Fruc}||\text{gluc6P}|} = \frac{\left(\frac{.0485}{\text{V}}\right)}{\left(\frac{.0485}{\text{V}}\right)^2} = 0.05$$

Solving for V gives a final solution volume of 78 mL.

PHASE DISTRIBUTION EQUILIBRIA

It often happens that two immiscible liquid phases are in contact, one of which contains a solute. How will the solute tend to distribute itself between the two phases? One's first thought might be that some of the solute will migrate from one phase into the other until it is distributed equally between the two phases, since this would correspond to the maximum dispersion (randomness) of the solute.

This, however, does not take into the account the differing solubilities the solute might have in the two liquids; if such a difference does exist, the solute will preferentially migrate into the phase in which it is more soluble.

For a solute S distributed between two phases a and b the process $S_a = S_b$ is defined by the distribution law,

$$K_{ab} = \frac{[S]_a}{[S]_b}$$

in which $K_{a,b}$ is the *distribution ratio* (also called the distribution coefficient) and $[S]_i$ is the solubility of the solute in the phase.

The transport of substances between different phases is of immense importance in such diverse fields as pharmacology and environmental science. For example, if a drug is to pass from the aqueous phase with the stomach into the bloodstream, it must pass through the lipid (oil–like) phase of the epithelial cells that line the digestive tract. Similarly, a pollutant such as a pesticide residue that is more soluble in oil than in water will be preferentially taken up and retained by marine organism, especially fish, whose bodies contain more oil-like substances; this is basically the mechanism whereby such residues as DDT can undergo *biomagnification* as they become more concentrated at higher levels within the food chain. For this reason, environmental regulations now require that oil-water distribution ratios be established for any new chemical likely to find its way into natural waters. The standard "oil" phase that is almost universally used is octanol, $C_8H_{17}OH$.

In preparative chemistry it is frequently necessary to recover a desired product present in a reaction mixture by extracting it into another liquid in which it is more soluble than the unwanted substances. On the laboratory scale this operation is carried out in a separatory funnel as shown here.

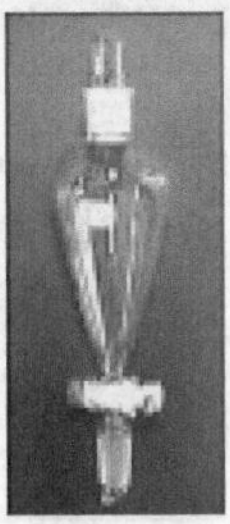

The two immiscible liquids are poured into the funnel through the opening at the top. The funnel is then shaken to bring the two phases into intimate contact, and then set aside to allow the two liquids to separate into layers, which are then separated by allowing the more dense liquid to exit through the stopcock at the bottom. If the distribution ratio is too low to achieve efficient separation in a single step, it can be repeated; there are automated devices that can carry out hundreds of successive extractions, each yielding a product of higher purity. In these applications our goal is to exploit the LeChâtelier principle by repeatedly upsetting the phase distribution equilibrium that would result if two phases were to remain in permanent contact.

Problem Example

The distribution ratio for iodine between water and carbon disulfide is 650. Calculate the concentration of I_2 remaining in the aqueous phase after 50.0 mL of 0.10M I_2 in water is shaken with 10.0 mL of CS_2.

Solution:

The equilibrium constant is,

$$K_d = \frac{C_{CS_2}}{C_{H_2O}} = 650$$

Let m_1 and m_2 represent the numbers of millimoles of solute in the water and CS_2 layers, respectively. K_d can then be written as $(m_2/10\text{ mL}) \div (m_1/50\text{ mL}) = 650$. The number of moles of solute is (50 mL) × (0.10 mmol mL^{-1}) = 5.00 mmol, and mass conservation requires that $m_1 + m_2 = 5.00$ mmol, so $m_2 = (5.00–m_1)$ mmol and we now have only the single unknown m_1. The equilibrium constant then becomes

$((5.00–m_1)$ mmol/ 10 mL) ÷ (m_1 mmol/ 50 mL) = 650.

Simplifying and solving for m_1 yields $(0.50–0.1)m_1/ (0.02\ m_1) = 650$, with$m_1$ = 0.0382 mmol. The concentration of solute in the water layer is (0.0382 mmol)/ (50 mL) = 0.000763 M, showing that almost all of the iodine has moved into the CS_2 layer.

The six Problem Examples presented above were carefully selected to span the range of problem types that students enrolled in first–year college chemistry courses are expected to be able to deal with. If you are able to reproduce these solutions on your own, you should be well prepared on this topic.

The first step in the solution of all but the simplest equilibrium problems is to sketch out a table showing for each component the *initial* concentration or pressure, the *change* in this quantity (for example, +2*x*), and the *equilibrium*values (for example.0036 + 2*x*). In doing so, the sequence of calculations required to get to the answer usually becomes apparent.

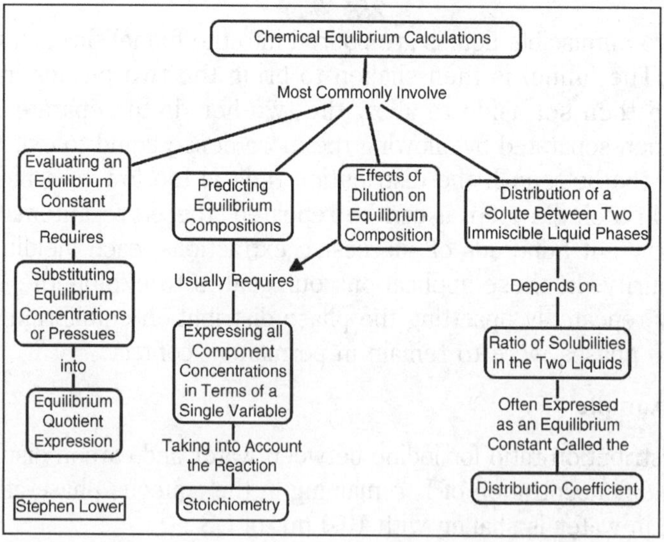

Equilibrium calculations often involve quadratic–or higher-order equations. Because concentrations, pressures, and equilibrium constants are seldom known to a precision of more than a few significant figures, there is no need to seek exact solutions. Iterative approximations are adequate and convenient.

Phase distribution equilibria play an important role in chemical separation processes on both laboratory and industrial scales. They are also involved in the movement of chemicals between different parts of the environment, and in the bioconcentration of pollutants in the food chain.

5

Environmental Soil and Water Chemistry

- Rates of chemical reactions depend upon the structures and concentrations of the reactants and products; the presence or absence of a catalyst; and the temperature.
- Reaction rates increase exponentially with temperature, as molecular kinetic energy increases according to the Maxwell-Boltzmann Distribution.
- Reaction rates increase in the presence of appropriate catalysts.

SOME BASIC IDEAS

Until now, our major concern has been whether or not a chemical reaction of interest may or may not occur under specified conditions of temperature, pressure, and concentration. The answer to this question is in the realm of thermodynamics. If thermodynamics tells us that a reaction may occur, then we must be concerned with how rapidly and by what pathway it will occur, and how we may influence this. The speed (rate) at which a reaction takes place is a matter of *chemical dynamics*, or alternately, *chemical kinetics*.

Chemical dynamics (kinetics) is the study of the rates (speeds) and mechanisms (pathways) of chemical reactions. We have briefly encountered the rate concept before in, when we discussed the manner in which vapour pressure changes with time; and in, when we presented dynamic equilibrium as a situation of balanced, opposing rates. Reaction rates will be our chief concern in this chapter. Although chemical kinetics is a well-developed field experimentally, we are still a long way from the theoretical grasp of rate that we have of equilibrium. Thus our predictive ability in the realm of kinetics is very limited.

Thermodynamics may predict that a reaction should proceed essentially completely. Whether it does so depends on the rate with which it occurs. For example, consider the reaction below at 298 K:

$$C(graph) + O_2(g) \rightarrow CO_2(g)\ [\Delta G^o_R = -394.4\ kJ]$$

Thermodynamics predicts that the reaction should go to completion. But carbon (coal) deposits exist in the atmosphere without noticeably reacting; and the "lead" in our pencils does not burst spontaneously into flame. The reaction is favoured, but is very slow at 298 K. We say that C(graph) is thermodynamically unstable in the presence of $O_2(g)$ at 298 K, but is kinetically inert. As a second case, consider the acid-base neutralization reaction:

$$H_3O^+(aq) + OH–(aq) \rightarrow 2H_2O(l)\ [\Delta G^o{}_R = –58.28\ kJ]$$

Thermodynamics predicts that this reaction, too, should go to completion. In fact, it occurs almost instantaneously; equilibrium is attained as rapidly as the reactants are mixed. Thus we say that H_3O^+ is both thermodynamically unstable and kinetically labile in the presence of OH^-. Two sets of descriptive words are necessary to distinguish thermo-dynamics and kinetics:

Thermodynamics: stable $\Delta G > 0$; unstable $\Delta G < 0$

Kinetics: inert (reaction is slow); labile (reaction is fast)

As we have seen above, there is no necessary correlation between "stable" and "inert", or "unstable" and "labile". This lack of correlation reflects the fact that thermodynamics deals with values of state functions in various equilibrium states of a system; kinetics deals with the rate at which a path between states is traversed. Reaction rates are extremely important in many ways in our lives. Sustained muscle action depends on the continuous supply of energy produced by glucose combustion in the mitochondria of cells. Energy production must be fairly rapid, otherwise sustained action is not possible. The extremely rapid (explosive) burning of gasoline in the internal combustion engine provides the motive power for automobiles.

Food cooks (or, as the chemist would say, oxidizes) extremely slowly at room temperature, but in a relatively short time at more elevated temperature. Our first concern must be to discuss the manner in which reaction rates are expressed, and to introduce some of the terminology of chemical dynamics.

What is a *rate*? In general, a rate is a number expressing how much a quantity changes in a convenient interval of time. Thus an interest rate is the amount by which a sum of money increases in 1 year; travel rate in an automobile or airplane is the distance covered in 1 hour; heart rate is the number of heartbeats per minute; thus a rate is the change in a quantity per unit time:

$$\text{Rate} = \Delta\text{quantity}/\Delta t$$

Note that the units of rate must be "quantity/time, or "quantity–time^{-1}". Thus interest rate has units of dollars/year; travel rate, miles/hour; and heart rate; beats/minute.

How are we to express the rate of a chemical reaction? To develop the answer to this question, we will begin with the simple chemical process in equation 15–1–2. The equation describes quantitatively the conversion of ozone to oxygen that occurs in the earth's upper atmosphere.

$$2O_3(g) \rightarrow 3O_2(g)$$

When this (or any other) reaction takes place, it is the amounts of reactants and products that change. As we know, amount can be expressed in either mass, moles, or molecules. The coefficients in the chemical equation relate both molecules and moles of reactants and products but do not directly relate masses. So we will agree that our measure of amount should be either molecules or moles. Further, since work in the laboratory is carried out on the macroscopic scale, moles is the more suitable unit of amount. So we might describe the rate of process 15–1–2 either in terms of the number of moles of O_3 used or of O_2 produced per unit time. However, there is a major difficulty with this approach. The number of moles of ozone used will depend upon the volume of atmosphere that we choose to examine!

We cannot specify a rate without also specifying the volume examined. Rather than use moles/time, then, we choose to divide by the appropriate volume to give moles/Litre-time, or concentration/time. Rates expressed in these units are independent of the actual amount of reaction carried out. At this point then we can express the rate of reaction 15–1–2 in either of two ways:

(15–1–3a): Rate = Moles O_3 used per Litre per time = –Δ(conc O_3)/Dtime

(15–1–3b): Rate = Moles O_2 produced per Litre per time = Δ(conc O_2)/Dtime

The units of the rate are then moles per Litre per time, where the time unit should be the one most convenient for the reaction being measured. Chemists agree that the rate of a reaction should be expressed as a positive number. As reaction 15–1–2 proceeds, the concentration of ozone becomes smaller while that of oxygen becomes larger. Thus the change in ozone concentration, Δ (conc O_3), is a negative quantity. To produce a positive rate in 15–1–3a, it is necessary to place the minus sign preceding the Δ (conc) term, as has been done.

In general, whenever the rate of a reaction is expressed in terms of the change in concentration of a reactant,D(conc reactant) is negative and the minus sign is necessary. The minus sign is unnecessary in 15–1–3b because Δ (conc O_2) is positive.

There is an additional difficulty, though. We would like to specify a single number representing the rate of a reaction under given conditions. However, the two expressions above give different numbers for the rate of 15–1–2, because dioxygen appears 1.5 times faster than ozone disappears (*i.e.*, in the same time that 2 moles of ozone react, 3 moles of dioxygen appear). This can be fixed by normalizing the two expressions in 15-1-3 with the stoichiometric coefficients. Dividing D(conc O_3) by the coefficient 2 gives the same number as dividing D (conc O_2) by the coefficient 3:

(15-1-4): Rate = –1/2 * Δ(conc O_3)/ time = 1/3 * D(conc O_2)/D time

Replacing "conc O_3" with the brackets to represent molar concentratration gives 15-1-5:

(15–1–5): Rate = –1/2 * Δ $[O_3]$/Dt = 1/3 * Δ $[O_2]$/D t

The symbol, Δ, signifies a finite (*i.e.*, measurable) change in a quantity. In the limit as the length of the time interval, Δt, approaches zero, we replace Δ(conc) and Δt with the differential (infinitesimal) quantities d(conc) and dt to obtain the *differential expression of rate* in 15-1-6.

(15–1–6): Rate = –1/2 Δ$[O_3]$/dt = 1/3 Δ$[O_2]$/dt

As we shall see in section 15–3, 15–1–5 is the appropriate formulation for finite difference methods of analysis.

For the completely general chemical reaction in 15–1–7, proceeding from left to right from reactants, A and B, to products, D and F, the rate of reaction can be represented using any one of the expressions in 15-1-8:

(15–1–7): aA + bB → dD + fF

(15–1–8): Rate = –1/a d[A]/dt = –1/b d[B]/dt = 1/d d[D]/dt = 1/f d[F]/dt

As discussed earlier, negative signs are necessary when rate is expressed in terms of disappearance of a reactant; and division by the stoichiometric coefficient guarantees that the same numerical value of rate is obtained from all four expressions.

Example. In the reaction of ozone to produce oxygen, it is found that under certain conditions of temperature and concentration, 0.0360 moles of ozone per Litre react in a 2–hour period. How much dioxygen is produced in this time period? What is the average rate of reaction over this time period?

Solution. The chemical equation, 15–1–2, tells us that 3 moles of O_2 are produced for each 2 moles of ozone reacted. Therefore 0.0360 moles ozone * (3 moles O_2/2 moles O_3) = 0.0540 moles dioxygen are produced.

The average rate of reaction can be expressed in either of two ways:

Rate = –1/2 Δ$[O_3]$/Δt = –1/2 (–0.0360 moles/L)/2 hours
= 0.009 moles/L–hour

Rate = –1/3 Δ$[O_2]$/Δt = 1/3 (0.0540 moles/L)/2 hours
= 0.009 moles/L-hour

Both expressions give the same positive rate of reaction.

Example: The rate of 15–1–9 was studied at 55°C by measuring the concentration of t–butyl bromide (t–BuBr) as a function of time. The data acquired are given below. Use the data to estimate the rate of reaction 20 seconds after reaction was begun.

(15–1–9): t-BuBr + $H_2O \rightarrow$ t-BuOH + HBr

Time, s	[t-BuBr]
0	0.100
10	0.0876
20	0.0768
30	0.0672
40	0.0590
50	0.0517
60	0.0453
80	0.0348
100	0.0267
120	0.0205
180	0.0093
240	0.0042

We begin by plotting the data to obtain a picture of the manner in which the concentration of t-BuBr varies with time. A plot of [t–BuBr] versus time is given in Figure. The shape of the plot is typical of a rate process: the concentration of reactant changes very rapidly at first, but as time goes on, the change in concentration per unit time becomes less and less as the concentration of t–BuBr approaches its final (equilibrium) value. We have seen this characteristic time plot before in our discussions of vapour pressure and approach to chemical equilibrium. Most concentration-time plots for a reaction in progress have this same general appearance.

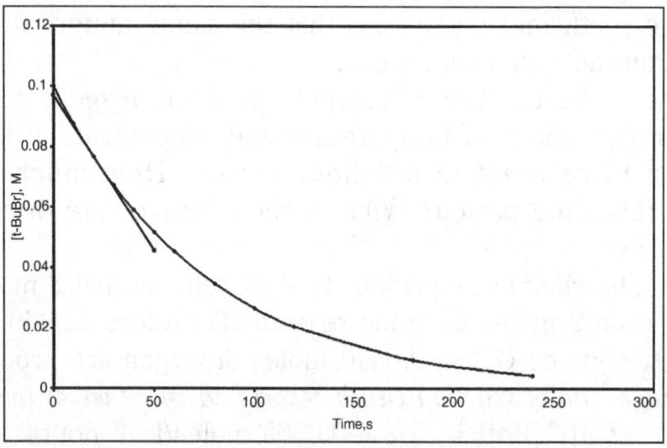

Fig. Time Profile of The Concentration of T-Butyl Bromide

Two approaches are possible for obtaining the rate at the 20-second mark of the reaction. The finite difference expression of rate in 15–1–5 can be used to calculate the average rate during the time interval preceding or following the 20-second mark. Both average rates are calculated below: Average rate between 10 and 20 seconds = –(conc at 20 s –conc at 10 s)/(20-10) Ms^{-1} = –(0.0768–0.0876)/(20-10) = 1.08 * 10^{-3} moles L^{-1} s^{-1}

Average rate between 20 and 30 seconds = –(0.0672–0.0768) (30-20) = $0.96 * 10^{-3}$ moles L^{-1} s^{-1}

The average rate over the 20–30 second time interval is less than that over the preceding interval because the reaction slows down as reactants are used up. This slowing down is readily apparent in the shape of the plot in Figure. However, the values are not too much different, so either may be used as an estimate of the rate of reaction at the 20 second mark.

The second approach is to use the differential expression of rate in 15-1-6. Just as dy/dx at x = x_o is the slope of a plot of y versus x at x =x_o, d[t-BuBr]/dt at a particular time, t_o, is the slope of a plot of [t-BuBr] versus time at the particular time, t_o. This slope is the same as the slope of the tangent to the curve at t_o. The tangent at t = 20 s is shown in the figure. It's slope is readily calculated to be 1.01×10^{-3} moles L^{-1} s^{-1}. As we might have expected, this value gives a rate intermediate between the values obtained above, based on finite time intervals. Because the slope of the tangent at a particular time represents the rate of reaction at that time, the rate obtained in this manner is called the *instantaneous rate* at the particular time. It is analogous to the speed that you measure when you glance at your automobile speedometer at a particular instant. As we have shown in this example, the instantaneous rate can be obtained by constructing tangents; or it can be estimated using the average rate over either the preceding or succeeding interval of time.

We can summarize our introductory discussion of reaction rate as follows:

- From a plot of [reactant] or [product] versus time, we can measure the rate at any particular time during the reaction as the slope of the tangent to the curve at that time; the initial rate is the slope of the tangent at t = 0.
- Rate decreases as the reaction proceeds towards equilibrium; this is evident from the shape of the concentration–time plot.
- The units of rate are []/time, or moles/L–time.

FACTORS AFFECTING REACTION RATE; THE RATE LAW

The rate of a chemical reaction depends on a number of factors, some of which are listed below:

- The molecular structure(s) of the reactant(s)
- The medium in which the reaction is carried out (for example, does the reaction occur in the gas phase, or does it occur with reactants dissolved in a solvent?)
- Temperature
- The presence or absence of a catalyst
- The concentrations of the reacting species

The dependence of rate on molecular structure is the most obvious but least understood factor. This complex matter is a focus of intense laboratory research. At present, it is not possible to summarize this dependence in a neat, clean package. In fact, this incompletely understood factor is what prevents predictive ability in kinetics. Consequently, we will postpone any discussion of this matter until a later chapter, in which the importance of chemistry in biology is discussed.

Similarly, the dependence of reaction rate on medium is a complicated matter that we will leave for more advanced books. Of the remaining three factors, temperature and the effects of catalysts are discussed in later sections of this chapter. In this section, we consider the effect of reactant concentrations on reaction rate.

For molecules to react, they must directly interact; to directly interact, they must collide. This leads us to expect that rate should be proportional to the frequency of molecular collisions. In turn, the frequency of collisions should depend on the number of molecules per unit volume, which is the concentration. This concentration dependence of rate is indeed observed experimentally. For most chemical reactions, the dependence of reaction rate on the concentrations of the reacting species is surprisingly simple.

In almost all cases, the reaction rate is found to be proportional to the product of the concentrations of the reactants, each concentration raised to an integer power that is almost always 1 or 2. Thus for the general reaction 15–1–7, the rate law is expected to take the form in 15–2–1.

$$(15\text{–}2\text{–}1)\text{: Rate} = -1/a\ d[A]/dt = k\ [A]^x\ [B]^y$$

Equation-1 is called the *differential rate law* for 15–1–7, because it involves the ratio of differentials (derivative), d[A]/dt. The exponents, x and y, are integers, and for most reactions are either 1 or 2. It must be stressed that In most cases x and y are different from the stoichiometric coefficients, a and b; it is not valid to write the rate law for a reaction by raising the concentrations to the stoichiometric powers, unless the reaction is an *elementary process*. We will have more to say about elementary processes in a subsequent section. The integer, x, is called the *order of the reaction in reactant A*; y is the *order in B*. The sum, x + y, is called the *overall reaction order*. The reaction orders must be determined by experiment; they can not be written down simply by looking at the reaction stoichiometry. To illustrate this statement for a specific case, the reaction below, involving 31 moles of reactants, has the accompanying rate law. This very complex overall reaction has a simple rate law, which is first order in each reactant, and second order overall.

$$30C_2H_5OH + B_{10}H_{14} \rightarrow 10B(OC_2H_5)_3 + 22H_2(g)$$

$$\text{Rate} = k[C_2H_5OH][B_{10}H_{14}]$$

Finally, k in 15–2–1 (and in the expression just above) is the proportionality constant relating Rate to the concentration product. It is called the *rate constant* for the reaction.

The rate constant contains implicit within it information about the dependence of rate on molecular structure, reaction medium, and temperature. A word about the units of the rate constant is in order. Generally, k has whatever units are required to give the rate units of concentration/time. Example 15–3 shows how units for the rate constant may be determined for various rate laws.

Example. For each rate law below, state the order of the reaction in each reactant, and determine the units of the rate constant, k.

$$-d[A]/dt = k$$

$$-d[A]/dt = k[A]$$

$$-d[A]/dt = k[A]^2$$

$$-d[A]/dt = k[A][B]$$

$$-d[A]/dt = k[A]^2[B]$$

Solution.

Rate Law	Reaction Order	Units of k
$-d[A]/dt = k$	Zero order (no dependence)	k has units of rate, Moles L^{-1} concentration $time^{-1}$
$-d[A]/dt = k[A]$	First order in A	Moles/L-time = k(moles/L) so k has units of $time^{-1}$
$-d[A]/dt = k[A]^2$	Second order in A	Moles/L-time = $k(moles/L)^2$ so k has units L $moles^{-1}$ $time^{-1}$
$-d[A]/dt = k[A][B]$	First order in A First order in B Second order overall	Same as for the previous case
$-d[A]/dt = k[A]^2[B]$	Second order in A First order in B third order overall	Moles/L-time = $k(Moles/L)^3$ so k has units $L^2 moles^{-2}$ $time^{-1}$

SOME SPECIFIC EXAMPLES OF RATE LAWS. ZERO-ORDER REACTIONS

Chemical reactions having overall order zero are rare. However, evaporation of a pure liquid, discussed an example of a physical process with a zero-order rate law. Recall that the rate of evaporation of a liquid having exposed surface area, A, is just,

$$Rate = k*A$$

As long as the exposed surface area remains constant, we may rewrite this as,

$$Rate = k', \text{ where } k' = k*A$$

Since this rate is independent of concentration of liquid, the process is zero order in liquid.

FIRST-ORDER REACTIONS

Reactions and physical processes exhibiting first order rate laws are very common. *Condensation* is a first order physical process. The rate of condensation of vapour to liquid in a closed system is given by,

$$\text{Rate} = k^*P_{vap}$$

where P_{vap} is the vapour pressure of the substance. However, P is related to concentration in moles/L by the ideal gas law, leading to,

$$\text{Rate} = -d[\text{vapour}]/dt = k^*[\text{vapour}]$$

Molecular rearrangements (often called isomerization reactions) frequently follow first order rate laws. Vision is critically dependent upon,

$$\text{cis-retinal} \rightarrow \text{trans-retinal}$$

where cis- and trans- retinal contain the same atoms, bonded together in the same way, but with somewhat different spatial arrangements. The two forms are shown in Figure. The rate of conversion of cis- to trans- form is given.

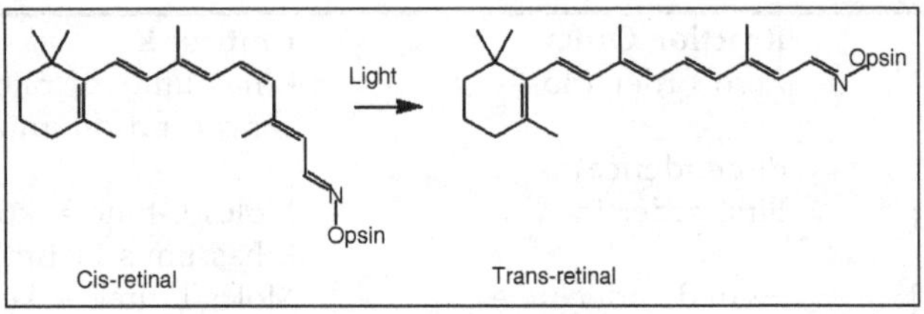

$$-d[\text{cis}]/dt = k[\text{cis}]$$

Similarly, trans- converts to cis- by a first order rate law with a different rate constant. Finally, nuclear decay processes are first order. Thus ^{238}U decays according to 15-2-8, which is governed by the rate law in 15–2–9.

$$(15\text{–}2\text{–}8)\text{: } {}^{238}\text{U} \rightarrow {}^{234}\text{Th} + {}^{4}\text{He}$$

$$(15\text{–}2\text{–}9)\text{: Rate} = -d(\text{mass } {}^{238}\text{U})/dt = k\ (\text{mass } {}^{238}\text{U})$$

The rate constant for this process has the extremely small value, 1.54×10^{-10} year^{-1}. The rate constant, k, for a first-order process has units of time^{-1}. It's reciprocal, 1/k, thus has units of time. The quantity, 1/k, is given the symbol t, and is called the*time constant* for the reaction. This is a measure of the time required for the reaction to take place. The time constant for decay of ^{238}U is $1/k = 6.49 \times 10^9$ years. The fact that this time constant is so large is the reason why ^{238}U is still found in the earth's crust, which is estimated to be about 4.5 billion years old.

Example: The time constant for the radioactive isotope, ^{226}Ra, is 2338 years. Calculate the number of atoms of ^{226}Ra that decay each minute in a 2.00 g sample of ^{226}Ra.

Solution: Nuclear decay is a first order process. We can therefore write the following rate law for decay of ^{226}Ra:

$$\text{Rate} = -d(^{226}\text{Ra atoms})/dt = k(\text{Number of } ^{226}\text{Ra atoms})$$

$$-dN/dt = kN$$

where N is the number of ^{226}Ra atoms in the sample, and $k = 1/\tau = 4.28 \times 10^{-4}$ year^{-1}.

$$N = (2.00\text{g}/226.025 \text{ g/mole}) * 6.023 * 10^{23} \text{ atoms/mole} = 5.33 * 10^{21} \text{ atoms.}$$

$$-dN/dt = (4.28 * 10^{-4} \text{ year}^{-1})(5.33 * 10^{21} \text{ atoms}) = 2.28 * 10^{18} \text{ atoms/year}$$

There are $5.256 * 10^5$ minutes in 1 year. Therefore,

$$-dN/dt = (2.28 * 10^{18} \text{atoms/year})/(5.256 * 10^5 \text{ minutes/year}) = 4.34 * 10^{12} \text{ atoms/minute}$$

SECOND ORDER PROCESSES

Many reactions exhibit second-order rate laws, either *simple* (second-order in a single reactant) or *mixed* (first-order in each of two reactants). Here we consider two specific reaction types that exhibit second-order kinetics. *Dimerization reactions,* in which two molecules unite to form a larger molecule, obey simple second-order kinetics. The dimerization of NO_2 is shown in 15–2–10.

$$(15\text{–}2\text{–}10)\text{: } 2NO_2(g) \rightarrow N_2O_4(g)$$

The rate law for 15–2–10 is 15-2-11:

$$(15\text{–}2\text{–}11)\text{: } -d[NO_2]/dt = k[NO_2]^2$$

where the second order rate constant is—$M^{-1}s^{-1}$ at °C. Other examples of dimerization processes are the formation of a Cl_2 molecule from two Cl atoms and the formation of one molecule of C_4H_8 from two molecules of C_2H_4. In general, a reaction of the general type in 15–2–12 is expected to obey simple second-order kinetics.

$$(15\text{–}2\text{–}12)\text{: } 2\,A \rightarrow A_2$$

Adduct formation processes, such as Lewis acid-base reactions and the formation of complexes involving enzymes and their substrates, are also second-order processes. Most such processes occur very rapidly. The role of enzyme-substrate complexes in enzyme catalysis will be discussed in a subsequent section.

"NET" REACTION RATE; A PRELUDE TO THE NEXT SECTION

In the next section we discuss experimental approaches to the determination of rate laws for reactions. To preface that discussion, we briefly discuss here the concept of *net reaction rate*. Except in some cases, we can experimentally measure only the net rate of a reaction, where the net rate is defined as in 15–2–13.

(15–2–13): Net reaction rate = rate of forward reaction – rate of reverse reaction

$$Rate_{net} = Rate_{forward} - Rate_{reverse}$$

The general reaction is reproduced below, this time acknowledging that in general the reverse process also occurs:

(15–1–7): $aA + bB \Leftrightarrow dD + fF$

For this reaction we expect the forward rate to depend only on the concentrations of reactants, A and B, and the reverse rate to depend only on the concentrations of products, D and F, as shown in 15–2–14:

(15–2–14a): $Rate_{forward} = 1\ k_f[A]^x[B]^y$

(15–2–14b): $Rate_{reverse} = k_r[D]^p[F]^q$

The net rate is therefore expected to be a potentially complex function of the concentrations of reactants AND products. The experimentally determined rate constant is a composite function of the forward and reverse rate constants, k_f and k_r.

(15–2–15): $Rate_{net} = k_f[A]^x[B]^y - k_r[D]^p[F]^q$

Once reaction reaches equilibrium, the net rate will of course be zero. Applying this idea to 15–2–13 leads to 15–2–16, our familiar criterion for dynamic equilibrium:

(15–2–16a): $Rate_{net} = 0 = Rate_f - Rate_r$

(15–2–16b): $Rate_f = Rate_r$

There are a few situations in which the net rate is equal to the forward reaction rate:

- One of the products escapes from the reaction system, so that its concentration remains at or very near zero at all times. A gaseous product that escapes from a solution is an example of this situation.
- The reaction has a large value of K_{eq}, so that the reverse rate is negligible except near equilibrium, which is not achieved until almost all of the limiting reactant has been used.
- At the very beginning of a reaction in which only reactants are initially present. At time zero, the reverse rate must be zero because no products are present yet.

EXPERIMENTAL DETERMINATION OF THE RATE LAW. THE INITIAL RATE METHOD

When applicable, the method of initial rates is a simple and clean way to obtain the rate law. The initial rate of a reaction is the instantaneous rate at t=0, when reactants are first mixed together and begin to react. In a plot of reactant concentration versus time, it is given by the slope of the tangent at t = 0. The slope of this type of plot is always steepest at the beginning; the initial rate is thus the fastest rate that we observe during the time course of the reaction process. One major advantage of the initial rate method is that the reverse rate is zero, provided only reactants and NO products are initially present. Thus only the forward rate contributes to the initial rate, and complications resulting from the reverse reaction are avoided. *The initial rate law depends only on concentrations of reactants*. The method of initial rates is best illustrated by example.

Example. The initial rate was measured at several combinations of initial reactant concentration for the following reaction:

$$A + 2B + 3C \rightarrow D + E$$

The rate law is expected to have the following general form:

$$d[A]/dt = k[A]^x[B]^y[C]^z$$

The following data were obtained. Use the data to determine the values of x, y and z in 15–3–2.

Run Number	$[A]_o$, M	$[B]_o$, M	$[C]_o$, M	Initial Rate, M/s
1	0.05	0.10	0.10	6.00×10^{-3}
2	0.05	0.15	0.10	13.5×10^{-3}
3	0.05	0.15	0.05	13.5×10^{-3}
4	0.10	0.10	0.05	12.0×10^{-3}

Solution. The key to the initial rate method is to *examine the dependence of rate on one reactant at a time, using data obtained when the other reactants do not vary in initial concentration*. In the first two data entries above, the initial concentrations of both A and C are the same in both runs; only the initial concentration of B is changed. These two data items can therefore be used to determine the order of the reaction in B. We see that when the initial concentration of B is increased from 0.10 to 0.15, an increase of a factor of 1.5, the rate increases from 6 to 13.5×10^{-3} moles/L-s, which is a factor of 2.25. We translate this information into equation form:

$$\text{Rate 2/Rate 1} = [B]_2^y/[B]_1^y = ([B]_2/[B]_1)^y \; 2.25 = (1.5)^y$$

We recognize that y must be 2. The reaction is therefore second order in B, and the rate law thus far is,

$$-d[A]/dt = k[A]^x[C]^z[B]^2$$

In runs 2 and 3, the initial concentrations of A and B are the same, while that of C is changed. We see from these data that even though the concentration of C is only half as large in run 3 as in run 2, the rate is unaffected. We conclude that the rate does not depend on [C], hence that z = 0. The rate law to this point is,

$$-d[A]/dt = k[A]^x[B]^2$$

There are no two kinetics runs in which only [A] is varied. However, runs 1 and 4 can be used to obtain the order in A because we have just determined that rate does not depend on [C]. Runs 1 and 4 indicate that the rate doubles when the initial concentration of A doubles. The order in A is thus 1, and the final rate law is,

$$-d[A]/dt = k[A][B]^2$$

Once the rate law is known, the rate constant may be calculated from any one of the individual runs. Choosing the first run,

$$k = Rate/[A][B]^2 = 6.00 * 10^{-3}/(0.05)(0.10)^2 = 12.0\ L^2/mole^2s$$

As this example shows, the initial rate method, ***when it is applicable***, provides a clean, simple path to the rate law. The italicized qualifier is important, however; in practice, the method of initial rates is only rarely used, for several reasons. In most kinetics studies, the chemist examines the manner in which the concentration of a selected reagent varies with time. The resulting data are presented either in a table, or in a plot of concentration versus time, as in Example and Figure. The data obtained at the beginning of the reaction, where the rate is largest and the time interval is smallest, is usually less reliable than that obtained at later times, when the rate is less. It is undesirable to base an analysis on one's least reliable data. Second, and more important, the method of initial rates uses only a small fraction of the concentration time data; the rest is wasted.

Third, to apply the initial rate method as was done in the example above requires that full concentration-time profiles be collected for each run, in order to extract the slope of the concentration-time plot at t=0.

Collection of this much data requires a lot of experimental effort. Most of the resulting data is then ignored if the initial rate method is used. For these reasons and others, chemists have developed other methods for determining rate laws from data, which make use of all of the data collected. Some of these will be discussed now.

INTEGRATION OF THE RATE LAW

It is possible to convert the differential rate law into an integrated form, in which concentration is expressed directly as a function of time. The concentration-time data obtained during a kinetics run in the lab may then be matched against the function. If the data conform to the function, it may be concluded that the reaction follows the corresponding differential rate law. Integration is very easy for rate laws of the form

$$-d[A]/dt = k[A]^n$$

We will examine the first and second–order cases, with n = 1 and 2 respectively. The first order rate law may be rearranged to put the concentration terms on one side and time on the other side of the equals sign, as in,

$$-d[A]/[A] = kdt$$

Both sides of the equation are then integrated over the limits, $[A] = [A]_o$ at t=0 and $[A] = [A]_t$ at some later time t:

$$\int_{[A]0}^{[A]t} d[A]/[A] = k\int_{0}^{t} dt$$

The integral of dx/x is ln x, and the integral of dx is x. Evaluating over the limits gives:

$$\ln([A]_t/[A]_o) = -k\,t$$

This is the so-called *integrated first order rate law*. For a reaction following simple first order kinetics, a plot of ln $[A]_t$ versus t is linear, with slope –k. Thus if concentration-time data for a reaction are plotted according to 15–3–9, and linearity is obtained, it may be concluded that the reaction follows a simple first-order differential rate law. An alternative expression of 15–3–9 is obtained by exponentiating both sides:

$$[A]_t = [A]_o e^{-kt}$$

Here e = 2.718... is the base of natural logarithms. Substituting the reciprocal of the time constant, t, for the rate constant gives,

$$[A]_t = [A]_o e^{-t/\tau}$$

Thus when an interval of time equal to the time constant has elapsed, the concentration of A will have decreased to 1/e (about 1/3) of its initial value. With the integrated first-order rate law in hand, we are in a position to consider a quantity called the *half-life*, $t_{1/2}$, which is similar but not identical to the time constant.

The half-life for a process is defined as the time required for the concentration of the limiting reactant to decrease to 1/2 of its initial value. An expression for the half life is easily obtained from 15–3–9 by realizing that $[A]_t = [A]_o/2$ at $t = t_{1/2}$. Making these substitutions in 15-3-9 leads to the following expression for the half-life of a first-order reaction:

$$t_{1/2} = \ln 2/k = 0.693/k$$

For n = 2 in corresponding to a second-order process, rearrangement leads to:

$$-d[A]/[A]^2 = k\,dt$$

Integration of both sides of this equation results in 15–3–11:

$$1[A]-1/[A]_o = k\,t$$

For a reaction with a simple second order rate law, a plot of concentration-time data as 1/[A] versus t is linear, with slope k. Thus a reaction for which the concentration-time data produce a linear 1/[A] versus t plot may be concluded to follow a second-order rate law.

Example: Plot the concentration-time data in Example according to equations and to determine whether the hydrolysis of t-butyl bromide obeys a simple first- or second-order rate law.

Solution: The concentration-time data from Example are reproduced below. Columns have been added for ln [t–BuBr] and 1/[t–BuBr]

Time, s	[t–BuBr], moles/L	ln [t–BuBr]	1/[tBuBr]
0	0.100	–2.303	10
10	0.0876	–2.435	11.42
20	0.0768	–2.567	13.02
30	0.0672	–2.700	14.88
40	0.059	–2.830	16.95
50	0.0517	–2.962	19.34
60	0.0453	–3.094	22.08
80	0.0348	–3.358	28.74
100	0.0267	–3.623	37.45
120	0.0205	–3.887	48.78
180	0.0093	–4.678	107.53
240	0.0042	–5.473	238.10

Figure is a plot of ln[t–BuBr] versus time; and Figure shows 1/[t–BuBr] against time. The linearity of the first plot indicates that the reaction is first–order in t–BuBr. The rate constant, obtained from the slope of Figure, is 1.32 * 10^{-2} s^{-1} (recall that for a first order reaction, the units of the rate constant must be $time^{-1}$, resulting from the division of rate in conc/time by concentration).

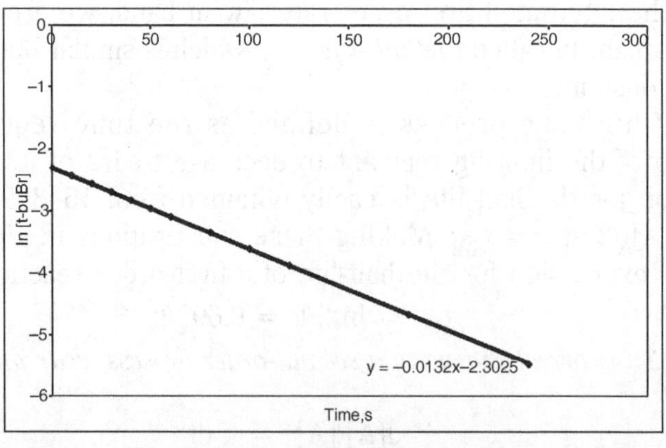

Fig. 10-3a: In [T-Bubr] Versus Time

Example: One product resulting from fission of ^{235}U in nuclear reactors is ^{137}Cs, a radioactive nucleus with a time constant of 43.3 years. How much time will it take for the radioactivity level of this waste to fall to 1% of its initial value, which is considered a safe level?

Solution: This problem is readily solved using equation 15-3-9:

$$\ln [A]/[A]_o = -kt = -t/\tau$$

$$t = -\ln(0.01)\tau = -\ln(0.01)(43.3) = 200 \text{ years}$$

The waste must be safely stored for 200 years before it will be safe.

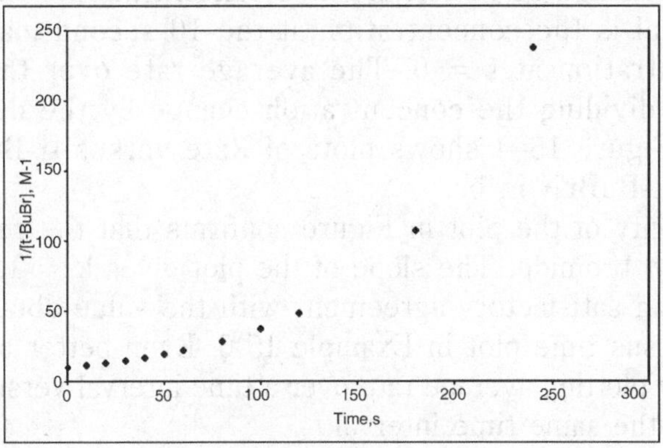

Fig. 10-3b: 1/[T-Bubr Vs. Time

FINITE DIFFERENCE METHODS OF NUMERICAL ANALYSIS

Another approach to determination of the rate law and rate constant from concentration-time data is to calculate the average rate over each experimental time interval. The average rates can then be plotted in turn against concentration, concentration squared, and so on until a linear plot is obtained. This is a relatively simple way to determine whether a reaction is first-order or simple second order (that is, has a rate law of the form, Rate = $k[A]^2$). Estimating derivatives (slopes) and integrals (areas) using finite changes in variables is called the Method of Finite Differences. We illustrate this method using the data for hydrolysis of t-BuBr from Examples 15–2 and 15–6.

Example: Use finite difference methods to obtain the rate law and rate constant for the hydrolysis of t-butyl bromide according.

Solution: The time and concentration data from Example 15–2 are reproduced below. In addition, values of Δt, Δ[t–BuBr], and rate = Δ[t–BuBr]/Δt have been calculated for each pair of successive data points.

Time, s	**Dt, s**	**[t–BuBr], moles/L**	**D[t–BuBr], moles/L**			**Rate * 10^3**	
0		0.100					
10	10	0.0876	–.0124	1.24			
20	10	0.0768	–.0108	1.08			
30	10	0.0672	–.0096	0.96			
40	10	0.0590	–.0082	0.82			
50	10	0.0517	–.0073	0.73			
60	10	0.0453	–.0064	0.64			
80	20	0.0348	–.0105	0.525	100	20	0.0267 –

.0081	0.40			
120	20	0.0205	–.0062	0.31
180	60	0.0093	–.0112	0.19
240	60	0.0042	–.0051	0.085

Thus the change in t-butyl bromide concentration over the first 10-second interval is the concentration at the 10 second mark minus the initial concentration at t = 0. The average rate over this interval is calculated by dividing the concentration change by the duration of the interval, Δt. Figure 15–4 shows plots of Rate versus [t-BuBr] in a and Rate versus $[t\text{-BuBr}]^2$ in b.

The linearity of the plot in Figure confirms that the reaction is first order in t-butyl bromide. The slope of the plot gives $k = 1.36 * 10^{-2}\ s^{-1}$. This value is in satisfactory agreement with the value obtained from the ln[t-BuBr] versus time plot in Example 15-6. Even better agreement can be achieved by plotting average rate over a time interval versus the average [t-BuBr] over the same time interval.

When the above data are plotted in this manner, the rate constant is found to be $1.33 * 10^{-2}\ s^{-1}$. Note that the method of finite differences, like the method based on integrated rate laws, uses the full range of concentration–time data, thus avoiding the major problem with the method of initial rates. For the method to be successful, however, it is important that the time intervals over which average rates are calculated be short relative to the total reaction time.

Example: The decomposition of N_2O occurs according to the following equation:

$$2N_2O(g) \rightarrow 2N_2(g) + O_2(g)$$

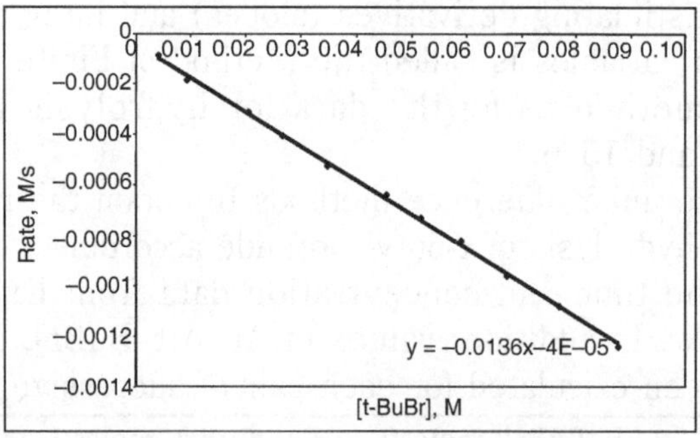

Fig. 10-4a: Rate Vs. [T-Bubr]

In the example above, the 10-second time intervals used in the early stages of the reaction, when concentration changes most rapidly, are small compared with the overall reaction time of 240 seconds. As a useful guideline, time intervals should not exceed 5% of the total reaction time; otherwise inaccurate rate constants will result from the finite difference approach. Electronic spreadsheets enable the calculations required for the finite difference method to be performed rapidly and conveniently using a computer. Further, the various data plots can be made quickly and professionally using the spreadsheet. With this modern tool, calculation by finite difference methods is routine. We will consider one additional example of the approach.

In a kinetics run, the concentration of N_2O was found to change with time as indicated in the first 2 columns of Table. The numbers in the remaining columns, calculated using an electronic spreadsheet, represent the average $[N_2O]$ over each time interval, squares of the N_2O concentrations, and the average rates of N_2O concentration change over the successive time intervals. Figure 10–4a and b show spreadsheet–generated plots of Rate versus $[N_2O]$ and Rate versus $[N_2O]^2$, respectively. Plot a shows slight but distinct curvature and in addition has a non–zero intercept. The linearity of the plot and lack of intercept in b confirms that the reaction follows the rate law in 15–3–13:

$$-d[N_2O]/dt = k[N_2O]^2$$

The rate constant, obtained from the slope of the plot, is $6.0 * 10^{-3}$ L/ mole-min.

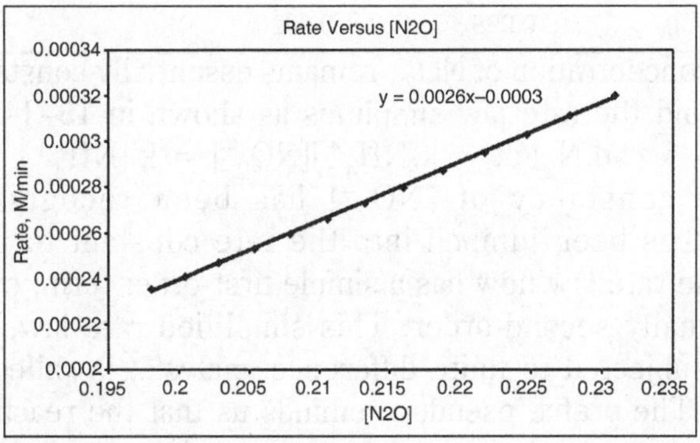

PSEUDO-ORDER METHODS

Many chemical reactions obey a mixed second-order rate law of the type shown:

$$\text{Rate} = k[A][B]$$

The reaction is first order in each of two reactants. An example of a reaction governed by this rate law is given in 15–3-15.

$$NH_4^+ + NO_2^- \rightarrow N_2(g) + 2H_2O$$

Suppose that a kinetics experiment is carried out by preparing a solution containing similar initial concentrations of the two reactants, NH_4^+ and NO_2^-, and monitoring the appearance of $N_2(g)$ as a function of time in order to measure the rate. Since the two reactants are present in similar concentrations, both substances undergo significant decreases in concentration over the course of the reaction. The rate of reaction reflects this simultaneous decrease in concentration of both reactants. Consequently, a plot of Rate versus $[NH_4^+]$ (NH_4^+ = A) will be non-linear, even though the reaction is first order in NH_4^+, because the Rate also responds to the change in concentration of NO_2^- (B) significantly during reaction. Although it is possible to integrate 15–3–14 to obtain a rather complex equation that may then be used to plot the data, a simpler approach to the kinetics study is often possible.In this approach, a kinetics run is performed on a solution that initially contains a substantial excess of one of the reactants, say B. If the initial concentration of B is more than 10 times larger than the initial concentration of A, the concentration of B will be nearly the same at the end of the reaction as at the beginning; consequently it may be considered to be constant during reaction. To illustrate this, consider a solution initially containing 0.002 M NH_4^+ and 0.10 M NO_2^-. If reaction is allowed to proceed until all NH_4^+ is consumed, the concentration of NO_2^- will have decreased by only 0.002 M to 0.098 M:

	NH_4^+ +	NO_2^- ®	N_2 +	$2H_2O$
Initial	0.002	0.10	0	
Change	–0.002	–0.002	0.002	
Final	0	0.098	0.002	

Thus the concentration of NO_2– remains essentially constant while NH_4^+ is consumed, and the rate law simplifies as shown in 15–3–16:

$$d[N_2]/dt = k[NH_4^+][NO_2^-] = k'[NH_4^+]$$

Here the constancy of $[NO_2^-]$ has been recognized, and this concentration has been lumped into the rate constant to produce a new constant k'. The rate law now has a simple first-order form, even though the reaction is actually second-order. This simplified rate law, which may be integrated or subjected to finite difference analysis, is called a *pseudo-first order* rate law. The prefix, pseudo-, reminds us that the reaction is not truly first-order, but has only been made to appear that way by imposition of a large concentration of NO_2^-. If one carries out kinetic runs using several (relatively large) concentrations of [B], a pseudo-first order rate constant, k', will be obtained for each run. Of course, these will have different values depending on [B]. A plot of k' versus [B] is then made to obtain the true rate constant, k, from the slope.

Example. Pseudo-First Order Kinetics. For the reaction of NH_4^+ with NO_2^-, 15–3–15, several kinetics runs were carried out using the initial reactant concentrations in the first two columns of the table below. Plots of Rate versus $[NH_4^+]$ for each run were found to be linear, indicating that the reaction is first-order in NH_4^+. The first-order rate constants obtained from these plots varied with $[NO_2^-]$ as indicated in the third column of the table. What is the order of the reaction in NO_2^-? What is the true rate constant for the reaction?

$[NH_4^+]$Initial	$[NO_2^-]$Initial	Observed 1st Order Rate Constant, k'
0.00896	0.197	$7.06 * 10^{-5}$
0.00904	0.395	$1.42 * 10^{-4}$
0.00523	0.593	$2.13 * 10^{-4}$

Solution. The order of reaction with respect to NH_4^+ is known to be 1, based on linear Rate vs. $[NH_4^+]$ plots. We thus focus on the relationship between the observed pseudo-first-order rate constants and the concentration of $[NO_2^-]$. Note that for all three runs, the initial NO_2^- concentration is much larger than the initial $[NH_4^+]$, guaranteeing that $[NO_2^-]$ remains essentially constant during the entire course of reaction.

A linear plot of k' versus $[NO_2^-]$ raised to a power, n, indicates that the reaction is n-order in $[NO_2^-]$. We can determine n by trying several plots, first with n = 1, then, if necessary, with n = 2, 3, and so on, until we obtain linearity. Plots of k' versus $[NO_2^-]$ and $[NO_2-]^2$ are shown in Figure.

The top plot is linear, so the reaction is first order in $[NO_2^-]$. The overall second-order rate constant, $k = 3.6 * 10^{-4}$ L/mole-s, is obtained from the slope.

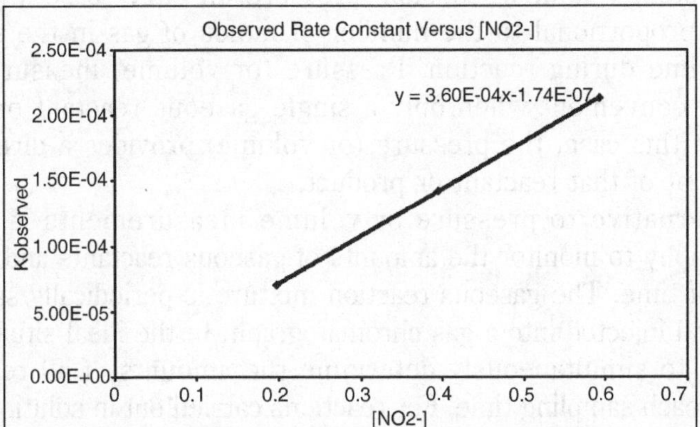

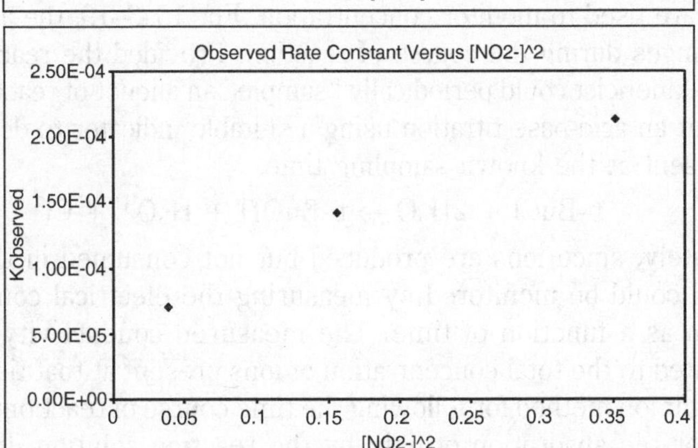

THE EXPERIMENTAL MEASUREMENT OF CONCENTRATION

In order to carry out a kinetics study, it is necessary to have an experimental "handle" on at least one of the reactants or products; that is, the change in concentration of a reactant or product must be accompanied by a change in some measurable property of the system. For reactions involving gases, the total pressure of the reaction mixture changes as reaction proceeds, provided that there is a net change in the amount of gas in the system. Thus, the rate of reaction 15-3-17 could be studied by monitoring the total pressure with time. Total pressure decreases as reaction proceeds because only 1 mole of gas forms for each 2 moles consumed.

$$2C_4H_6 \rightarrow C_8H_{12}$$

On the other hand, no change in total pressure accompanies 15-3-18 because there is no change in moles of gas; reaction rate can not be studied by monitoring total pressure in this case.

$$H_2(g) + I_2(g) \rightarrow 2HI(g)$$

Rather than measuring system pressure at constant volume, one could measure system volume at constant pressure. In either case, the measured quantity is proportional to the number of moles of gas in the system at a particular time during reaction. Pressure (or volume) measurements are particularly convenient when only a single gaseous reactant or product is involved. In this case, the pressure (or volume) provides a direct measure of the amount of that reactant or product.

An alternative to pressure or volume measurements is to use gas chromatography to monitor the amounts of gaseous reactants and products as a function of time. The gaseous reaction mixture is periodically sampled with a syringe and injected into a gas chromatograph. In the ideal situation, it may be possible to simultaneously determine the amounts of all reactants and products at each sampling time. For reactions carried out in solution, a number of methods are used to monitor concentration. For 15–3–19, the acidity of the solution changes during the course of reaction. Provided the reaction is slow enough, the kineticist could periodically "sample" an aliquot of reaction solution and carry out an acid-base titration using a suitable indicator to determine the $[H_3O^+]$ present at the known sampling time.

$$t\text{–}BuCl + 2H_2O \rightarrow t\text{–}BuOH + H_3O^+ + Cl^-$$

Alternately, since ions are produced but not consumed in the reaction, the kinetics could be monitored by measuring the electrical conductivity of the solution as a function of time. The measured conductivity values can then be related to the total concentration of ions present at that time. Perhaps the most common method for following the time course of reactions in solution is to monitor the absorption of light by the reacting solution. If one of the reactants (or products) has an absorption band in the UV–visible, infrared, or NMR region of the electromagnetic spectrum, the change with time of the amount of light absorbed is a measure of the rate of appearance or disappearance of the absorbing substance. The kinetics of 15–3–20 can easily be studied by this method.

$$Cr(H_2O)_6^{3+} + EDTA^{4-} \rightarrow Cr(EDTA)^- + 6H_2O$$

The product, $Cr(EDTA)^-$, has an absorption maximum at a wavelength of 542 nm in the visible region of the spectrum. As reaction proceeds, absorbance at this wavelength increases in direct proportion to the concentration of $Cr(EDTA)^-$, providing a "handle" on the reaction kinetics.

ELEMENTARY PROCESSES AND REACTION MECHANISM

A balanced chemical equation is intended to represent the *net* transformation of reactants to products that occurs when reaction takes place. It does not, and is not intended to, tell us anything about the details of the reaction pathway. In fact, most reactions occur via a series of relatively simple steps, each involving reaction of only a few (one, two, and very occasionally, three) molecules. Each of these simple steps is called an *elementary process*. The equation for an elementary process represents an actual collision event at the molecular level. The sequence of elementary processes by which a reaction is proposed to occur is called the *mechanism* for the reaction. For example, the reaction in 15–4–1 is thought to occur by the 2-step mechanism in 15–4–2a and 15–4–2b.

Overall:

$$(15\text{–}4\text{–}1):\ 2NO_2(g) + F_2(g) \rightarrow 2NO_2F(g)$$

Mechanism:

$$(15\text{–}4\text{–}2a):\ NO_2 + F_2 \rightarrow NO_2F + F$$

$$(15\text{–}4\text{–}2b):\ NO_2 + F \rightarrow NO_2F$$

with the first step proposed to be slow, the second one fast. Each of the two steps in the mechanism is understood to be an elementary process. We can make several important generalizations about the mechanism in 15–4–2. First, each elementary process involves reaction (collision) of only two species, even though the overall reaction involves 3 moles of reactants. Second, F atoms, produced in the first step and consumed in the second, do not appear in the overall equation for the reaction. Such a species—first produced, then consumed—is called a *reaction intermediate*. Third, the elementary processes add to give the overall equation for the reaction. This must always be true. It is important to realise that a mechanism for a reaction is a hypothesis about what goes on at the molecular level, made by the chemist based on her or his experimental observations. The overall equation for a reaction is NOT a hypothesis; it is an experimental fact, observable in the laboratory. The mechanism in all of its detail is seldom observable; it must be deduced from experiment, must be consistent with all known experimental facts about the reaction, and is subject to change if in conflict with a new experimental fact.

TYPES OF ELEMENTARY PROCESSES

There are three types of elementary process, named for their *molecularity*—the number of reactant molecules that they involve. A *unimolecular process* involves only one reactant molecule. Two examples of unimolecular processes are given.

$$O_3 \rightarrow O_2 + O$$

$$I_2 \rightarrow 2I$$

A *bimolecular process* involves two reactant molecules, which may be the same or different. Examples are given. Bimolecular elementary processes are by far the most common type.

$$NO_2 + F \rightarrow NO_2F$$

$$H_2O_2 + I^- \rightarrow H_2O + IO^-$$

A *termolecular process* requires the simultaneous collision of three reactant molecules. is an example.

$$2NO + O_2 \rightarrow N_2O_4$$

Such three-body collisions are extremely improbable. Consequently, termolecular processes are rare. Elementary processes with molecularity greater than three do not occur, due to the overwhelming improbability of collisions involving four or more molecules. In contrast to the situation for the overall (net) reaction, *the rate law for an elementary process may be written directly from its stoichiometry*, because an elementary process is an actual, not a net, process. Thus for the elementary process,

$$A + B \rightarrow C + D$$

the rate law is $-d[A]/dt = k[A][B]$. Unimolecular processes have first-order rate laws. For, the rate law is

$$-d[O_3]/dt = k[O_3]$$

Bimolecular processes have rate laws that are second-order overall; and termolecular processes have overall third order rate laws.

$$-d[NO_2]/dt = k[NO_2][F]$$

$$-d[O_2]/dt = k[NO]^2[O_2]$$

In summary, an elementary process of molecularity n has a rate law of order n. The converse is NOT true: a first-order reaction is not necessarily either an elementary process or unimolecular. The difference between order and molecularity is subtle but important. The concept of molecularity is appropriate only in the context of an elementary process, and indicates the number of reactant molecules involved in (undergoing collision) in the elementary step. The order of a reaction is the number to which the concentration of a reactant must be raised in the experimental rate law for the overall (net) reaction. This rate law, and the orders in it, do not necessarily correspond to a particular elementary step in the mechanism of the net reaction. We will say more about this shortly.

RATE DETERMINING STEP

A particular overall reaction consists of a sequence of elementary steps, which usually occur with different rates. Very often one of the elementary processes occurs much more slowly than do the others and limits the rate at which the overall reaction occurs. This slow step is then referred to as the*rate-determining step (rds). The overall rate is equal to the rate of the rds*. The fact that the overall reaction rate is limited by the rate of a particular elementary step is the reason why rate laws frequently do not correspond with the stoichiometry of the overall reaction. A familiar example of an overall process occuring at a rate limited by a slow step is illustrated in Figure. Pictured is a theater filled to capacity, connected by a set of wide double doors to a smaller lobby, which is linked to the outside via a revolving door. When the movie is over, people begin to move from the theater to the lobby, and from there to the outside. We suppose that the double doors between theater and lobby are large enough so that 10 people per second can pass through; but that only 1 person per second can exit the revolving door. We then represent the overall process of leaving the theater in terms of 2 steps:

$$\text{theater} \rightarrow \text{lobby [10/s]}$$
$$\text{lobby} \rightarrow \text{outside [1/s]}$$
$$\text{Overall, theater} \rightarrow \text{outside [Rate = ?]}$$

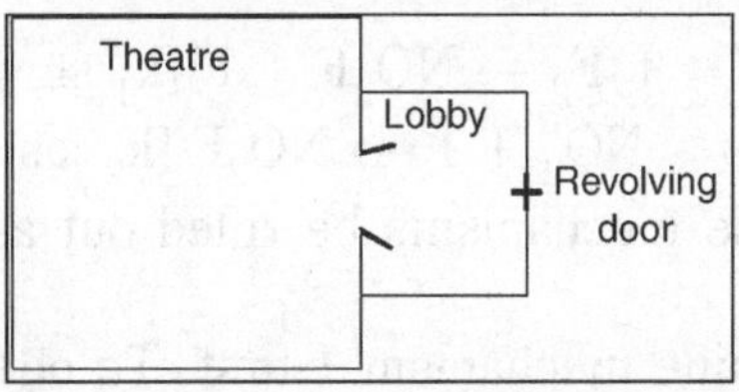

Initially, people pass from theater to lobby at the rate of 10 per second, and begin to exit the revolving door one by one. However, the lobby quickly fills up, because it cannot be emptied nearly as rapidly as it was filled. The process "piles up" behind the slow step, which is passage through the revolving door. The maximum rate at which people can leave the theater is 1 per second, which is the rate of the slowest step in the process.

RATE LAW AND REACTION MECHANISM

Chemists are very interested in the details of how a reaction occurs; *i.e.*, in the reaction mechanism.

The approach to the elucidation of the mechanism of a reaction of known stoichiometry involves at least the following steps:

- Do experiments to determine the experimental rate law for the overall reaction;
- Propose a mechanism for the overall reaction, based on the observed rate law;
- Derive the theoretical rate law implied by the proposed mechanism;
- Compare the rate laws in 1 and 3. If they have the same form, it is safe to conclude that the mechanism is consistent with the experimental rate law. If they do not have the same form, go back to 2.

Step 2 makes use of the fact that the *observed rate law indicates what atoms are involved as reactants in the rate-determining step of the mechanism.* This will be made clear in specific examples to follow. It is the most difficult step of the above procedure, requiring extensive experience and well-developed "chemical intuition." At this level of study, we will not attempt to deduce mechanism from the rate law. Rather, our focus will be this: given the form of the experimental rate law, and one or more proposed mechanisms, decide whether each mechanism is or is not consistent with experiment.

Example: For the reaction $2NO_2(g) + F_2(g) \rightarrow 2NO_2F(g)$, the experimental rate law has the form, Rate = $k_{obs}[NO_2][F_2]$. Two mechanisms have been proposed. These are given below:

Mechanism 1

$$NO_2 + F_2 \Leftrightarrow NO_2F + F \; [k_1, k_{-1} \text{ fast}]$$
$$NO_2 + F \rightarrow NO_2F \; [k_2 \text{ slow (rds)}]$$

Mechanism 2

$$NO_2 + F_2 \rightarrow NO_2F + F \; [k_1 \text{ slow (rds)}]$$
$$NO_2 + F \rightarrow NO_2F \; [k_2 \text{ fast}]$$

Can either of these mechanisms be ruled out as being in conflict with experiment?

Solution: We examine mechanism 1 first. To obtain the theoretical rate law for this mechanism, we proceed as follows:

1. Write the rate law for the rds. The rate of this step is the same as the rate of the overall reaction.

The rds for mechanism 1 is step 2.

$$\text{Rate} = k_2[NO_2][F]$$

Although this rate law is valid, it is not expressed solely in terms of concentrations of the reactants in the overall reaction. Since these concentrations are the ones we actually observe, we must somehow rewrite the rate law so as to eliminate the concentration of the intermediate, F, which is not generally observable, and replace it with concentrations of reactants in the net reaction. We can accomplish this making use of the preceding rapid steps, which we assume to occur rapidly enough to come to a pseudo equilibrium. So,

2. Use the rapid pre-equilibrium to solve for the concentration of the intermediate in terms of known concentrations. Since the first step of the mechanism is in equilibrium, the forward and reverse rates of this elementary process are the same, and we can write

$$k_1[NO_2][F_2] = k_{-1}]NO_2F][F]$$

Solving for [F] gives,

$$[F] = k_1/k_{-1}\ \{[NO_2][F_2]/[NO_2F]\}$$

This expression for the concentration of the intermediate can be substituted into the rate law for the rds:

$$\text{Rate} = k_1k_2/k_{-1}\ \{[NO_2]^2[F_2]/[NO_2F]\}$$

This is the rate law implied by the mechanism. It is the rate law that the reaction should follow if the proposed mechanism is correct. Clearly, though, it is not in agreement with the experimental rate law, which is first order in NO_2, and shows no dependence on NO_2F. The conclusion is that Mechanism 1 can NOT be correct; it is in conflict with experiment.

We next consider Mechanism 2, using the same general approach:

1. Write the rate law for the rds, which in this case is the first step in the mechanism. Rate = $k_1[NO_2][F_2]$ Further work is unnecessary here, because the rds rate law is already expressed in terms of concentrations of participants in the overall reaction. The implied rate law for Mechanism
2 Has the same form as the experimental rate law, and is therefore not in conflict with experiment.

Can we conclude, then, that Mechanism 2 is the correct mechanism for reaction 15–4–1? The answer is an emphatic NO. If a mechanism does not agree with experiment, it must be wrong; but just because a mechanism agrees with experiment does not mean that it is correct. A mechanism is a theory—it can never be proved. To underscore this point, consider the following possible mechanism for 15–4–1:

Mechanism 3

$$NO_2 + F_2 \rightarrow NO_2F_2\ [k_1 \text{ slow}]$$

$$NO_2F_2 + NO_2 \rightarrow 2NO_2F\ [k_2 \text{ fast}]$$

This also is consistent with experiment, giving the implied rate law

$$\text{Rate} = k[NO_2][F_2]$$

based on the first step, which is the slow one. You might like to assess the validity of yet a fourth possible mechanism, given below:

Mechanism 4

$$1/2\ F_2 \rightarrow F\ [\text{fast}]$$

$$F + NO_2 \rightarrow NO_2F \text{ [slow]}$$

We can now appreciate the significance of a statement made earlier: the experimental rate law tells us what atoms are contained in the reactants of the rate-determining step. The experimental rate law in Example 15–11, Rate = k $[NO_2][F_2]$, reveals that the reactants in the slow step must contain a total of 1 N atom, 2 O atoms, and 2 F atoms. Based on this, we can immediately rule out mechanism 1, for which the rds involves only one F atom; and mechanism 4, for which the same is true. Both mechanisms 2 and 3 have rds's containing the correct numbers of atoms, so neither is in conflict with experiment.

In general, chemists try to propose mechanisms that agree with the experimental rate law and with any other experimental evidence; and that are chemically reasonable. It is the assessment of chemical reasonability that requires experience. We consider one additional example of mechanism assessment.

Example. Several mechanisms might be considered for the overall reaction

$$2NO + O_2 \rightarrow 2NO_2$$

with rate law, Rate = $-d[O_2]/dt = k_{obs}[NO]^2[O_2]$. The symbol, k_{obs}, where "obs" stands for "observed", is commonly used for an experimentally measured rate constant.

Mechanism 1:

$$NO + NO \Leftrightarrow N_2O_2 \text{ [}K_1 \text{ fast]}$$

$$N_2O_2 + O_2 \rightarrow 2NO_2 \text{ [}k_2 \text{ slow]}$$

Mechanism 2:

$$NO + O_2 \Leftrightarrow OONO \text{ [}K_1' \text{ fast]}$$

$$NO + OONO \rightarrow 2NO_2 \text{ [}k_2 \text{ slow]}$$

Mechanism 3:

$$NO + O_2 \rightarrow OONO \text{ [}k_1 \text{ slow]}$$

$$NO + OONO \rightarrow 2NO_2 \text{ [}k_2 \text{ fast]}$$

Can any of these mechanisms be ruled out?

Solution: Mechanism 3 can be eliminated right away, because the reactants in the rds do not contain the correct numbers of atoms. The experimental rate law demands that there be two N atoms and four O atoms in the rds reactants; however, only one N and three O atoms are involved in the rds of mechanism 3. Mechanisms 1 and 2 show the correct atoms in the rds. You should confirm that the implied rate laws are

$$\text{Mechanism 1: Rate} = k_2K_1[NO]^2[O_2]$$

$$\text{Mechanism 2: Rate} = k_2K_1'[NO]^2[O_2]$$

both of which are consistent with experiment. Attempts to draw Lewis structures for the intermediates, N_2O_2 and OONO, are shown in Figure. The intermediate OONO is a free radical (it has an odd electron), and should therefore react very rapidly with the free radical NO.
It is thus unlikely that the second step of Mechanism 2 would be slow. Mechanism 1 is preferable, based on chemical reasonability. There is yet a fourth mechanism, involving a one-step termolecular process. It is currently believed that this is the correct one.

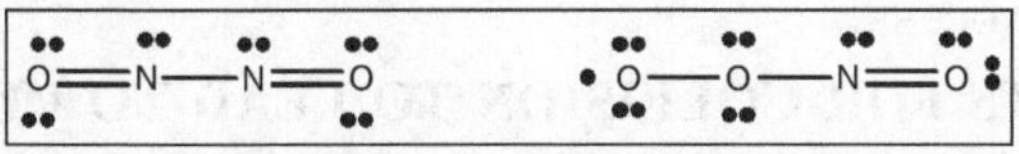

Let's summarize what we have just learned about reaction mechanisms:

- The steps in a mechanism are numbered from 1 to n. The subscripts on the elementary process rate constants are given the same numbers, with k_n being the forward rate constant and k_{-n} the reverse rate constant for step n. The steps must add to give the overall reaction.
- The mechanism often (but not always) involves at least one slow step. This is the rate-determining step.
- Any fast step preceding the slow step is treated as reversible, and assumed to come to equilibrium.
- To obtain the rate law, we
  - Write the rate law for the elementary rds;
  - Eliminate concentrations of intermediates using rapid pre–equilibrium expressions. This is necessary because the concentrations of intermediates are not usually observable.
- That the reaction proceeds in several steps having different rate constants can cause its rate law to deviate from the overall stoichiometry. However, the rate law tells us which atoms are involved in the reactants of the rate determining step, so is strongly suggestive of mechanism.
- Not all elementary step rate constants need appear in the implied rate law for the mechanism.
- Mechanisms can be disproven, but not proven.

TEMPERATURE DEPENDENCE OF RATE—KINETIC MOLECULAR THEORY

The rates of chemical reactions increase with increasing temperature, often dramatically. Coal reacts with oxygen at a negligible rate at room temperature, but burns spontaneously with the liberation of heat and light at 500 °C.

A sirloin steak "cooks" (oxidizes) very slowly at room temperature, but requires only 20 minutes at 180°C. Reactions are accelerated by temperature increase because the elementary processes by which they occur take place more rapidly.

We begin this section with two questions:

1. Why are some elementary processes fast and others slow at a particular temperature?
2. Why are reaction rates (rate constants) so dramatically temperature dependent?

We can arrive at answers to these questions by looking in more detail at the collision process.

REQUIREMENTS FOR COLLISION TO LEAD TO REACTION

As discussed earlier, rate is proportional to the number of molecular collisions occuring per unit volume per unit time. For the decomposition of hydrogen iodide to hydrogen and iodine, shown in 15–5–1, assuming that [HI] = 0.001 M and a temperature of 500 ºC, the rate would

$$2HI(g) \rightarrow H_2(g) + I_2(g)$$

be 5.8×10^4 M/s if all collisions led to formation of products. The observed rate, however, is only $1.2 * 10^{-3}$ M/s, $5 * 10^7$ times slower than it would be if all collisions between HI molecules were effective. Our conclusion must be that only a small fraction of molecular collisions lead to formation of products; *i.e.*, are*effective*. In general, for the elementary process,

$$A + B \rightarrow C + D$$

Rate = Number of effective collisions/volume-time, which should be proportional to the total number of collisions per unit volume per unit time. We call this total number of collisions Z. We can relate Z to the concentrations of A and B via equation 15–5–3:

$$Z = Z_0[A][B]$$

where Z_0 is the number of collisions per volume per time when A and B have 1 molar concentrations. The number of effective collisions is less than the total number of collisions for two reasons. *First, two colliding molecules must be properly oriented if they are to react.*

Reaction requires that bonds within reactants be broken and that new bonds form between atoms in the colliding molecules. A bond cannot form between atoms in separate molecules unless they are near each other during collision. Figure shows two of the many possible collision orientations of CO and NO_2, which are reactants.

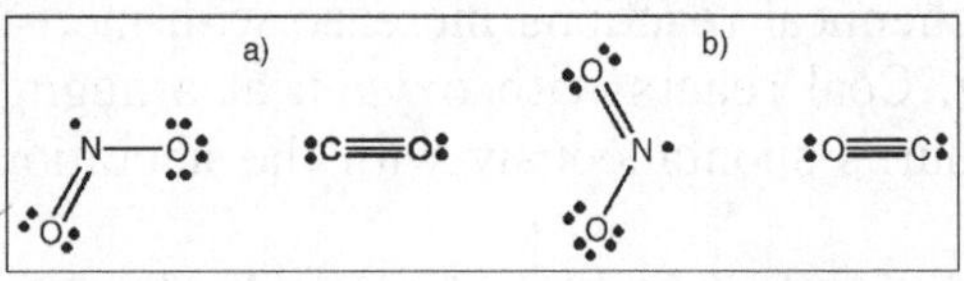

$$CO + NO_2 \rightarrow CO_2 + NO$$

This bimolecular elementary process involves transfer of an oxygen atom from NO_2 to CO. For this to occur, it is necessary that an oxygen atom of NO_2 approach the carbon atom of CO during collision, so that a bond may form between them. This requirement is satisfied in the orientation in Figure. The relative orientation in brings the nitrogen atom of NO_2 and the oxygen atom of CO together, so cannot be effective in advancing the process. Similarly, many other possible orientations of these two molecules are unsuitable for transfer of an oxygen atom from N to C, so will not be effective. We symbolize the fraction of collisions having the proper orientation of molecules as p, and recognize that p is a small number.

Second, for reaction to occur, it is necessary that colliding molecules have some minimum total kinetic energy. This is called the activation energy, symbolize E_a. In almost all cases, successful reaction requires that existing bonds be broken and new bonds made. Two bonded atoms lie at a minimum in a potential energy curve of the type discussed and pictured in Figure.

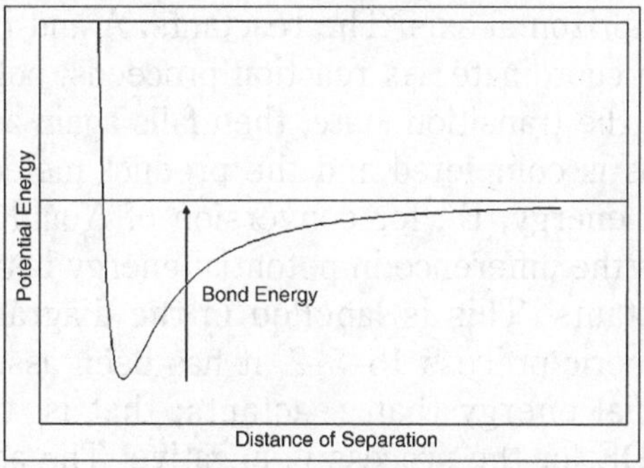

Fig. 10-9 Potential Well

For the bond to break, an input of energy at least equal to the depth of the well is required. This energy is drawn from the total kinetic energy possessed by the colliding species. If the kinetic energy is insufficient to overcome the bonding interactions, the reactant molecules separate without change. Even if the total kinetic energy is sufficient for reaction, the collision orientation may still be inappropriate for reaction.

Both factors—sufficient kinetic energy and correct orientation—are necessary for reaction to occur. Assuming that the minimum energy is available and the orientation correct, bonds will be ruptured, new bonds will form with the release of energy, and new molecules will separate from the collision center. Between reactants and products, both of which are relatively low in potential energy, lies a state of maximum potential energy called the *transition state*.

The arrangement of atoms that exists in this state is called the *activated complex*. We may think of it as a "molecule" in which chemical bonds are in flux, some being broken and others being made. We can depict this situation graphically in terms of a *reaction coordinate diagram*, which is a plot of potential energy versus the progress of reaction, which is usually called the "reaction coordinate." A reaction coordinate diagram for the bimolecular elementary process shown in Figure.

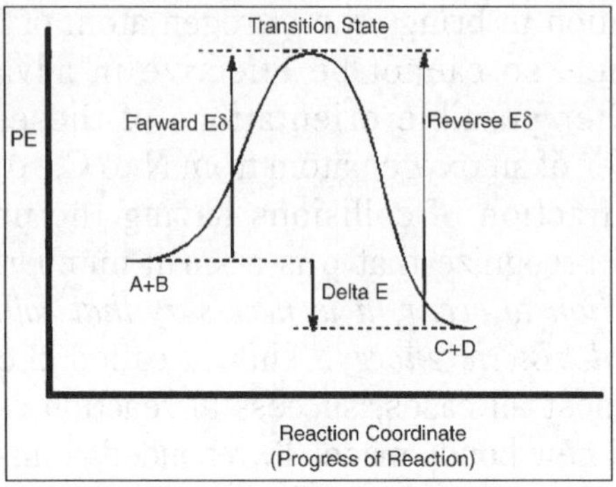

Potential energy is plotted on the vertical axis, with the reaction coordinate on the horizontal axis. The reactants, A and B are located to the left on the reaction coordinate. As reaction proceeds, potential energy rises to the maximum at the transition state, then falls again as bond formation in the product species is completed and the product molecules separate.

The activation energy, E_a, for conversion of A and B to C and D (the forward reaction) is the difference in potential energy between the transition state and the reactants. This is labelled in the diagram. In drawing this diagram for the generic process 15–5–2, it has been assumed that products have lower potential energy than reactants; that is, that the process is exothermic. Thus DE for the process is negative. The activation energy for the reverse process, also indicated in the diagram, is the difference in PE between the transition state and the potential energy of C + D. It is clear from the figure that ΔE for the reaction is the difference between the activation energies for the forward and reverse processes. This is generally true.

$$\text{(15–5–5): } E_a(\text{forward}) = \text{PE(activated complex)} - \text{PE(A+B)}$$
$$E_a(\text{reverse}) = \text{PE(activated complex)} - \text{PE(C+D)}$$
$$E_a(\text{forward}) - E_a(\text{reverse}) = \Delta E \text{ for elementary process}$$

Every elementary process can be depicted as a single-humped potential energy curve, with an activated complex at the maximum. Letting the fraction of collisions with sufficient kinetic energy to react be f, then

$$\text{(15–5–6): Rate = Number of effective collisions/volume-time} = pfZ$$

We are now in a position to answer the first of the two questions posed at the beginning of this section. Some elementary processes are slower than others either because they have very restrictive orientation requirements, corresponding to a very small value for p; or they have high potential energy barriers (E_a), corresponding to a small value for f. Next we consider the effect of an increase in temperature. The distribution of kinetic energies in a collection of molecules is governed by the Maxwell-Boltzmann distribution, first introduced.

The M-B distribution curve for temperature T_1 is shown in Figure. The value of the activation energy for elementary process has been indicated on the graph. The fraction of molecules with kinetic energy greater than E_a is given by the area under the M-B curve to the right of E_a. As temperature is increased to some new value T_2, the M–B curve shifts to the right and flattens, indicating the increase in both the average molecular kinetic energy and the breadth of the distribution. The high temperature curve is also shown in the figure. Notice that the area under the curve to the right of E_a is much larger for the high-temperature curve, leading to a dramatic increase in the number of effective molecular collisions. Quantitatively, the fraction of effective collision can be expressed.

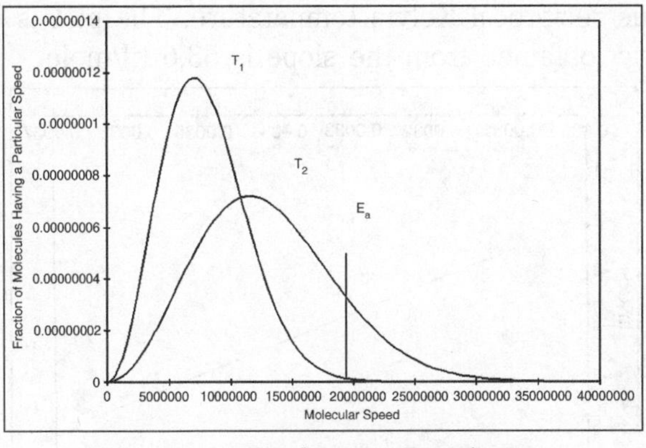

Fig. Maxwell-Boltzmann Distribution

$$f = \text{Area under curve right of } E_a \ \alpha\ e^{-Ea/RT}$$

Thus,

$$\text{Rate} = pfZ = pe^{-Ea/RT}Z_o[A][B]$$

Equation 15-5-8 has the form of a rate law if we identify the terms preceding the concentrations of A and B as the rate constant, k:

$$k = pZ_oe^{-Ea/RT} = Ae^{-Ea/RT}$$

where A contains information about the orientation requirements. Taking the logarithm of both sides of 15-5-9 gives 15-5-10, called the Arrhenius equation.

$$\ln k = \ln A - E_a/RT$$

The value of the activation energy for a process is obtained by measuring the rate constant at each of several temperatures, and constructing a plot of lnk versus 1/T. The plot should be linear, with slope = $-E_a/R$ and intercept = lnA. We are now able to offer an answer to the second of the two questions posed at the outset: reaction rates are temperature dependent because of the requirement of a minimum kinetic energy for effective collisions.

Example. Aqueous hydrogen peroxide and iodide ion react according to the following equation.

$$H_2O_2(aq) + 3I^-(aq) + 2H^+(aq) \rightarrow I_3^-(aq) + 2H_2O$$

The reaction is first order in peroxide and first order in iodide, second order overall. The reaction was studied at several temperatures, with the following results. What is the activation energy for the reaction?

T, °C	k_{obs},	$M^{-1}s^{-1}$
2	3.03 *	10^{-3}
17	1.0 *	10^{-2}
50	7.6 *	10^{-2}

Solution. The activation energy can be obtained from the slope of a plot of ln k_{obs} versus reciprocal Kelvin temperature. The plot is in Figure. The activation energy obtained from the slope is 53.6 kJ/mole.

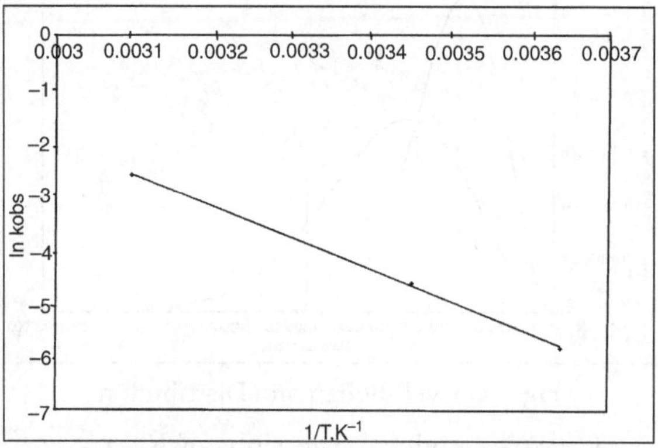

Fig. 10-13 Arrhenius Plot For Example 15-10

OVERALL (MULTISTEP) REACTIONS

Thus far we have discussed the temperature dependence of a single elementary process. Similar considerations apply to overall (net) reactions consisting of several steps.

We make several general statements about reaction coordinate diagrams for net reactions before looking at a specific example:

- The number of maxima in a reaction coordinate diagram is equal to the number of steps in the mechanism.
- The slowest step usually has the highest E_a.
- The rate law for the overall reaction indicates the atoms present in the activated complex for the slowest elementary step.

- Intermediates occur at minima in the reaction coordinate diagram.
- The E_a measured experimentally for the multistep reaction is the difference between the potential energy at the highest point in the diagram and the potential energy of the reactants.

Example: For the reaction

$$2NO_2 + F_2 \rightarrow 2NO_2F$$

the rate law is $-d[F_2]/dt = k_{obs}[NO_2][F_2]$. Propose a mechanism for the reaction, and draw an appropriate reaction coordinate diagram.

Solution: The rate law tells us that the activated complex of the slow step contains two oxygen atoms, two fluorine atoms, and one nitrogen atom. This suggests the following mechanism:

$$NO_2 + F_2 \rightarrow NO_2F + F \ [k_1 \text{ slow}]$$

$$F + NO_2 \rightarrow NO_2F \ [k_2 \text{ fast}]$$

Because reaction involves the breaking of the F-F bond and formation of a bond between N and F, the structure shown in Figurea is a reasonable arrangement of atoms in the activated complex. Finally,

Figure shows the reaction coordinate diagram corresponding to the proposed mechanism. There are two "humps" (maxima), one corresponding to each elementary step in the mechanism; the intermediate, F, is located at the minimum between maxima; the activation energy for the overall reaction is indicated; and the overall reaction is exothermic. The activation energy for the first step is substantially larger than that for the second step, consistent with the first step being rate-determining. Thus the reaction coordinate diagram reveals by its profile which is the slow step in the overall reaction.

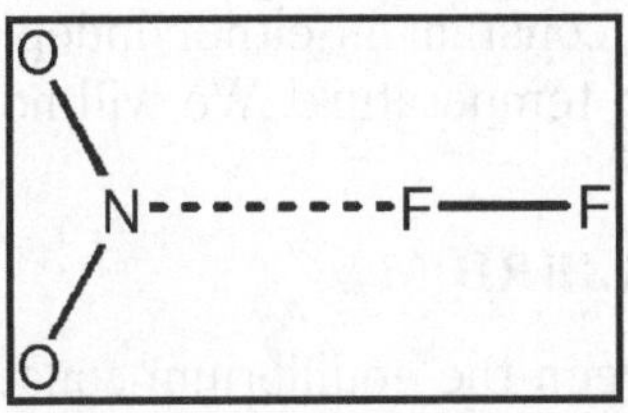

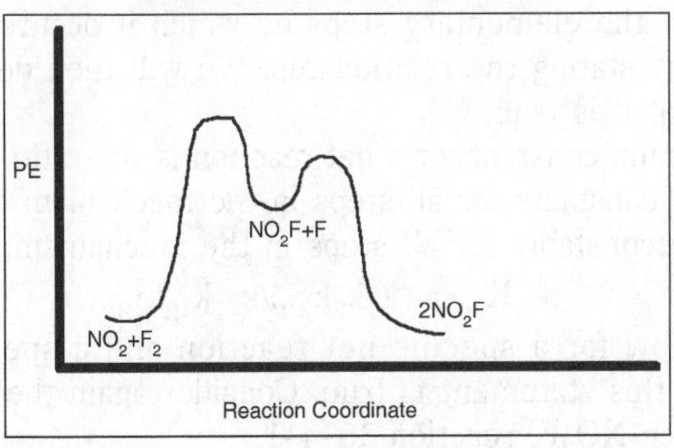

Example: The reaction coordinate diagram for an unspecified reaction is shown in Figure. From the details of the diagram, say as much as possible about the net reaction to which it corresponds.

Solution: Because there are three maxima in the diagram, the mechanism must consist of three steps. Step 2 must be the rds, because it has the largest E_a. There is one activated complex for each step, for a total of three. There are two minima, indicating two intermediates. The experimental rate law for the reaction should indicate the composition of the activated complex of the second step. The overall reaction is exothermic, because products are lower on the energy axis than reactants. Finally, the measured E_a is the energy difference between the highest maximum in the diagram and the reactants. It is shown in the diagram.

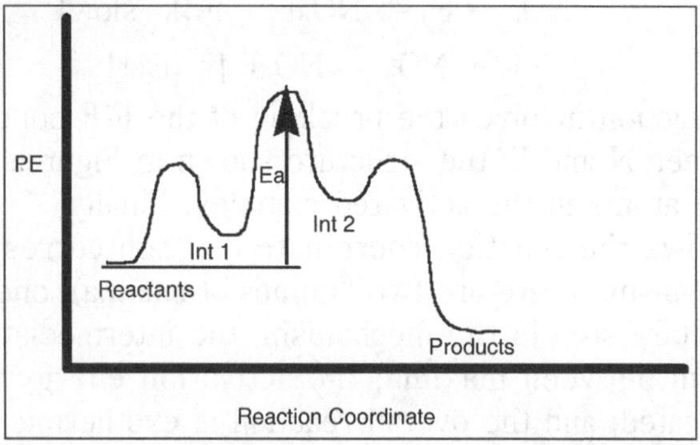

The temperature dependence of the observed rate law for an overall reaction consisting of several steps is usually qualitatively similar to that for a single elementary process. However, there are complex mechanisms for which this is not the case. Indeed, examples are known of reactions for which the observed rate constant is either independent of temperature or decreases with increasing temperature. We will not be concerned with such situations in this book.

KINETICS AND EQUILIBRIUM

The connection between the equilibrium constant for a reaction and the rate constants of the elementary steps by which it occurs is an important one. We begin by stating the relationship. We will then demonstrate, for a specific case, that it is true.

The equilibrium constant for a net reaction is the ratio of the product of the forward rate constants for all steps in the mechanism to the product of the reverse rate constants for all steps in the mechanism:

$$K_{eq} = k_1k_2k_3.../k_{-1}k_{-2}k_{-3}...$$

We now show for a specific net reaction and a specific (proposed) mechanism that this statement is true. Consider again the reaction of NO_2 and F_2 to produce NO_2F, reaction 15–4–1:

$$2NO_2 + F_2 \Leftrightarrow 2NO_2F$$

We have written the reaction with the double arrow because we are now interested in the forward and reverse rates at equilibrium. At equilibrium, the overall forward rate and overall reverse rate must be equal:

$$Rate_f = Rate_r$$

This must be true regardless of the detailed pathway by which the reaction occurs. All available experimental evidence indicates that the mechanism for 15–4–1 is the two-step process in 15–4–2a and b:

$$NO_2 + F_2 \Leftrightarrow NO_2F + F \; [k_1, k_{-1}]$$

$$NO_2 + F \rightarrow NO_2F \; [k_2, k_{-2}]$$

When the overall reaction is in equilibrium, so must each elementary step be. Consequently, we have used double arrows in the elementary steps as well. The equality of the forward and reverse rates of the individual steps is expressed mathematically as follows.

$$k_1[NO_2]_e[F_2]_e = k_{-1}[NO_2F]_e[F]_e$$

$$k_2[NO_2]_e[F]_e = k_{-2}[NO_2F]_e$$

Eliminating the concentration of the intermediate, F, gives.

$$[NO_2F]_e{}^2/[NO_2]_e[F_2]_e = k_1k_2/k_{-2}k_{-2} = K_{eq}$$

The statement made at the outset is thus seen to be true for this specific case: *The equilibrium constant is the ratio of the product of forward rate constants to the product of reverse rate constants*. For the general overall reaction occuring by an n-step mechanism,

$$K_{eq} = k_1k_2...k_n/k_{-1}k_{-2}...k_{-n}$$

Objection is frequently made to the method of derivation that we have just used, because the mechanism for is not known with certainty; it is and will remain a hypothesis, however well based in experiment. If it is found that the two step mechanism in 15–4–2 is indeed NOT correct, then the relationship in is invalid. This is certainly true. However, even if the mechanism above is incorrect, must occur by some mechanism.At equilibrium each step in the mechanism must be balanced in rate, enabling us to eliminate concentrations of any and all intermediates just as we did above, to arrive at a modified version of 15–6–2. The validity of the general relationship in 15–6–3 is independent of the details of mechanism.

Thus the intuitively-expected connection between k and K does indeed exist. In fact, there are many examples in which the equilibrium constant for a reaction has been obtained from kinetics studies. This process can never be reversed however; it is not possible to obtain k by performing equilibrium studies. Before leaving this matter, we make one more very important point. There is a tendency to believe that reactions with large equilibrium constants are fast, whereas those with small equilibrium constants are slow. However, neither of these beliefs is true. A large value for K_{eq} means that the product of forward rate constants is much larger than the product of reverse rate constants. It does not follow, however, that the forward rate constants are Large. It means only that they are larger than the reverse rate constants. $k_1k_2...k_n >> k_{-1}k_{-2}...k_{-n}$ does not mean that $k_1,k_2...,k_n$ are large in the absolute sense.

A large K_{eq} can result from the ratio of a slow forward rate to an even slower reverse rate.

CATALYSIS

A number of chemical reactions that ordinarily occur slowly can be induced to occur more rapidly by the addition of a suitable substance called a *catalyst*. We will define a catalyst as a substance that speeds up a reaction without itself being consumed or chemically changed in the overall reaction process.

This definition is in practice rather restrictive, because eventually, all real catalysts become deactivated by "poisoning" or via irreversible structural changes; thus they cannot function indefinitely. For our purposes, here, however, this definition will serve.

The reactant molecule that is affected by the catalyst is called the *substrate*. The process by which a catalyst effects its action is called *catalysis*. Catalysis is one of the most intensely studied areas in science because it is of tremendous biological and industrial importance. We begin our exploration of catalysts by discussing how they work.

THE MODE OF CATALYST FUNCTION

A catalyst functions either by entering into a slow step of the uncatalyzed reaction mechanism, or by creating an entirely new mechanism for the reaction. In either case, it is thought that the catalyzed pathway has a lower activation energy than the uncatalyzed path.

We can illustrate catalytic action generically in terms of the following reaction scheme, where A, B, D, and F are reactants and products in the overall reaction, and C is a catalyst for the reaction.

$$A + B \rightarrow D + F \text{ [overall]}$$

Uncatalyzed mechanism:

$$A + B \rightarrow AB \text{ [slow]}$$

$$\text{(15–7–2b): } AB \rightarrow D + F$$

Catalyzed mechanism:

$$A + C \rightarrow CA \text{ [fast]}$$

$$CA + B \rightarrow CAB$$

[less slow than first step of uncatalyzed mechanism]

$$CAB \rightarrow D + CF \text{ [fast]}$$

$$CF \rightarrow C + F \text{ [fast]}$$

The effect of the catalyst on the activation energy of the slow step of the reaction is shown schematically in Figure. As shown in the figure, the catalyst affects only E_a; it does not affect the energy of either the reactants or products of the reaction. The intermediate, CAB, in which both reactants are bound to the catalyst, is of particular interest.

A possible (and fairly common) structural motif for this intermediate is shown in Figure. Notice that, although A and B are bound to different sites on the catalyst, C, they are in proximity and can effectively interact. Binding of A and B to C causes shifts in electron density that may facilitate bond breaking within A and/or B and bond formation between A and B or fragments of them.

Thus the catalyst provides an organizing center for A and B, facilitating their interaction. In this manner the catalyst can overcome the orientation factor discussed earlier.

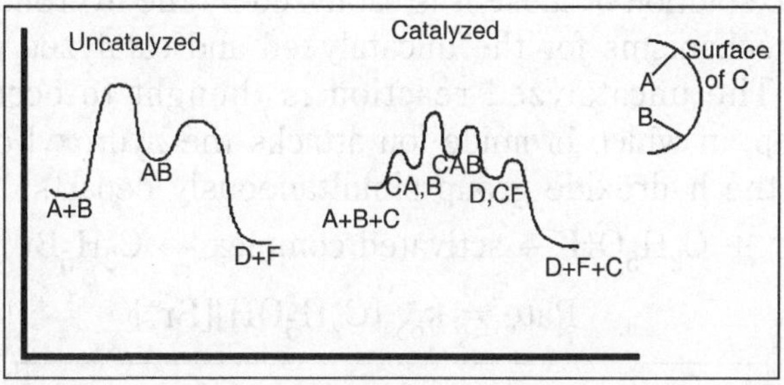

As soon as the molecule of catalyst is regenerated in the last step of the 4-step mechanism, it may bind another molecule of A and proceed once again through the sequence, converting A and B to D and F. This sequence is repeated a large number of times, so that generally only a small amount of catalyst is required to convert a substantial quantity of reactant to product. Chemists call the process by which the catalyst cycles through the reaction over and over again a catalytic loop, and represent it as in Figure. The unbound form of the catalyst, C, is placed at the 12 oclock position of the loop, and the various forms in which the catalyst is found are placed more or less evenly around the remainder of the loop.

Arrows show the direction of reaction around the loop, with reactants brought in from outside the loop, and products ejected out of the loop. This is a very effective visual presentation of catalyst action. The number of times that a molecule of catalyst cycles through the loop per unit time is called the *turnover number* of the catalyst. An effective measure of turnover number is moles product produced per mole catalyst present per time.

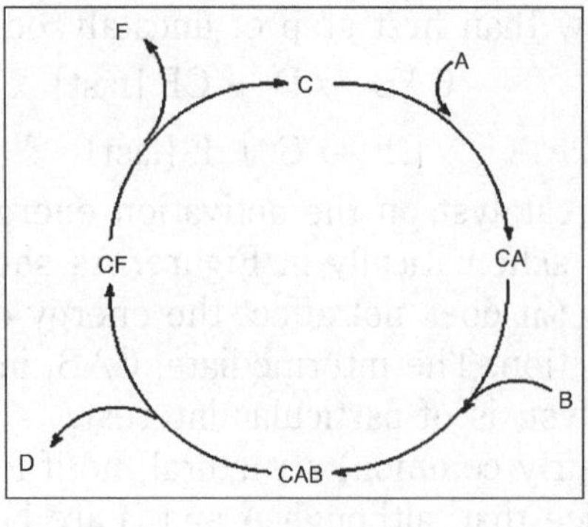

Let's look at a specific example of the effect of a catalyst on a simple reaction, from which we can draw some general conclusions.

Example: The reaction of ethanol (ethyl alcohol) with bromide ion to produce ethyl bromide and hydroxide ion is shown in 15–7–4.

$$C_2H_5OH + Br- \rightarrow C_2H_5Br + OH^-$$

The reaction is quite slow in neutral or basic solution, but occurs quite rapidly in acidic solution because it is catalyzed by the hydronium ion, H_3O^+. Discuss the mechanisms for the uncatalyzed and catalyzed processes.

Solution. The uncatalyzed reaction is thought to occur in a single elementary step, in which bromide ion attacks the hydroxyl carbon atom of ethanol while the hydroxide group simultaneously departs:

$$Br^- + C_2H_5OH \rightarrow \text{activated complex} \rightarrow C_2H_5Br + OH^-$$

$$\text{Rate} = k_{obs}[C_2H_5OH][Br^-]$$

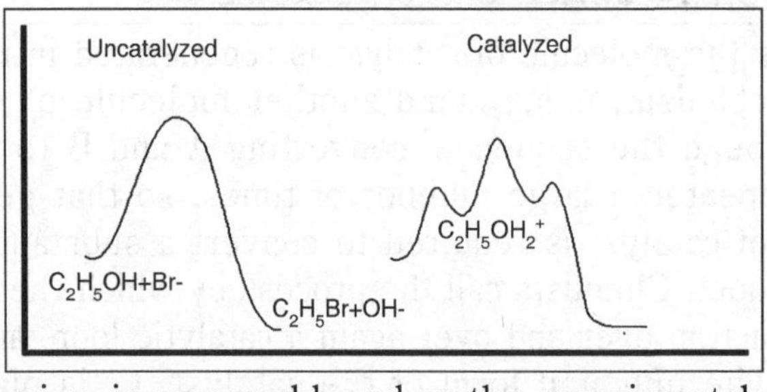

This mechanism is proposed based on the experimental rate law, which is overall second order. The observed rate law for the catalyzed process is

$$\text{Rate} = k_{obs}[Br^-][C_2H_5OH][H_3O^+]$$

The following three-step mechanism is consistent with this rate law if k_{obs} is identified with k_2K_1:

$$(15\text{–}7\text{–}6a): C_2H_5OH + H_3O^+ \rightarrow C_2H_5OH_2^+ + H_2O \text{ [fast, } K_1]$$

$$(15\text{–}7\text{–}6b): C_2H_5OH_2^+ + Br- \rightarrow C_2H_5Br + H_2O \text{ [slow } k_2]$$

$$(15\text{–}7\text{–}6c): 2H_2O \rightarrow H_3O^+ + OH^-$$

The reaction coordinate diagrams for the uncatalyzed and catalyzed pathways are shown in Figure.

Several general statements about catalysis are evident from this example:

- Because the catalyst does not appear in the overall equation for the reaction, it does not affect the equilibrium; it affects only the kinetics. The catalyst speeds up not only the forward reaction, but also the reverse, by the same factor. The position of equilibrium remains the same.
- The activation barrier is lower in the catalyzed mechanism. Even though the catalyzed path involves more steps, the overall reaction occurs more rapidly.
- The concentration of the catalyst appears in the rate law, even though the catalyst does not appear as a reactant or product in the overall equation.
- Although a catalyst and an intermediate share the property of not appearing in the overall equation, they are different types of things. First, the catalyst appears in the rate law; an intermediate cannot. Second, the catalyst appears first as a reactant, then is generated later as a product. An intermediate appears first as a product, then later as a reactant.

EXAMPLES OF CATALYSIS

We will now briefly discuss several examples of catalysis that are actually used on a mammoth scale in the chemical industry. Each of these processes, in its own way, has a major impact on the quality of our lives.

The Polymerization of Ethylene

Ethylene is a very simple molecule with formula C_2H_4. It is a gas, obtained as a byproduct during the catalytic cracking of petroleum. Under certain conditions, ethylene molecules can be made to join together end-to-end to form very long chain-like molecules of *polyethylene*, so called because it consists of m

$$(15-7-7): 2bCH_2 = CH_2 \xrightarrow{Ticl_3} -(CH_2\,CH_2\,CH_2\,CH_2)_{n^-}$$

Even at high temperature and pressure of ethylene, this process occurs negligibly slowly. In the presence of a small amount of a modified form of $TiCl_3$ (called tickle-3 in the plastics industry), however, it occurs rapidly at only moderately high temperature and pressure. The process is referred to as Ziegler-Natta catalysis after its two coinventors, who jointly received the Nobel Prize in Chemistry in 1963. Since the discovery of this process in the 1950's, the entire plastics industry has developed and grown to huge proportions. To this day, the mechanism of Ziegler-Natta catalysis is not fully understood. Mechanistic studies are difficult for several reasons, one of which is that the process is *heterogeneous*; that is, the catalyst ($TiCl_3$) and substrate (ethylene) are in different phases. The catalytic process takes place on the surface of crystals of $TiCl_3$, which rapidly become covered with and blocked from view by the resulting polyethylene. Much effort is ongoing in the US chemical industry to develop more efficient and easily handled Ziegler-Natta catalysts.

The Production of Sulfuric Acid

Year after year, sulfuric acid ranks first on the list of the top ten chemical substances produced in the United States: billions of pounds are produced annually. Sulfuric acid is synthesized by the so-called *Contact Process*, which involves the four sequential steps below:

$$S + O_2 \rightarrow SO_2 \text{ [fast]}$$

$SO_2 + O_2 \rightarrow SO_3$ [slow, because it occurs by a termolecular elementary process catalyzed by V_2O_5]

$$SO_3 + H_2SO_4 \rightarrow H_2S_2O_7 \text{ [fast]}$$

$$H_2S_2O_7 + H_2O \rightarrow 2H_2SO_4 \text{ [fast]}$$

The reaction of SO_3 with water is very exothermic and causes extensive spattering and production of a fine mist of highly acidic water. For this reason, direct reaction of SO_3 with water in the third step is impractical. Instead, SO_3 is bubbled into pure sulfuric acid, with which it reacts smoothly to give fuming sulfuric acid, $H_2S_2O_7$.

This can then be treated with the stoichiometrically correct amount of water to give sulfuric acid. The contact process has been so perfected that sulfuric acid is very inexpensive to produce. Consequently, it is used in any industrial process requiring acid. It finds it major uses in the production of phosphate fertilizers; in paper manufacture; in the petroleum industry; in steel production; and in the production of detergents. The importance of these products in our lives is obvious.

Catalytic Converters

For some years now, catalysts have been placed within the exhaust systems of automobiles to reduce the amount of poisonous or otherwise harmful emissions. These noble metal catalysts (based on platinum and palladium) carry out a dual function. First, they facilitate oxidation of carbon monoxide, resulting from incomplete hydrocarbon combustion, to carbon dioxide:

$$CO(g)+1/2O_2(g) \xrightarrow{Pt} CO_2(g)$$

Second, they catalyze decomposition of nitric oxide, produced during engine operation and oxidized rapidly to toxic NO_2 by atmospheric oxygen, to N_2 and O_2. Unfortunately, catalytic converters also facilitate oxidation of SO_2 to SO_3, which is the precursor of acid rain. Low-sulfur petroleum distillates are therefore essential. The Haber Process. Vast quantities of ammonia are synthesized each year for use as fertilizer. Currently the most efficient process for ammonia synthesis is the Haber-Bosch Process, developed during the ten-year period preceding 1913, in which nitrogen and hydrogen react directly at high temperature and pressure and in the presence of an activated iron catalyst to form ammonia.

$$N_2(g) + 3H_2(g) \xrightarrow{Fe} 2NH_3(g)$$

A catalyst and high temperature are necessary to cause the reaction to go at a reasonable rate. Unfortunately, high temperature makes the exothermic reaction less favoured, so very high pressure is used to favour products.

Even with conditions optimized, reaction is incomplete, and unreacted hydrogen gas is recycled for maximum efficiency of ammonia production. Nitrogen fertilizers produced from ammonia are largely responsible for the incredible growth in agricultural production over the decades since the First World War, when the process was first put on line in Germany. Ironically, the original motivation for development of the process was the requirement of explosives for the war effort.

The processes discussed above have at least three features in common. First, catalysis is heterogeneous; the catalyst is in all cases a solid, interacting with the substrate in the gas phase. Second, the mechanisms for these processes are incompletely understood. Thus catalysts are used successfully on a huge scale, even though we do not understand how they work. Third, in all cases the catalyst involves a transition metal, from the d block of the periodic table. Transition metals are often versatile catalysts because they are flexible in coordination number (that is, the number of atoms, ions, or molecules to which they may bind in a Lewis acid-base interaction); in stereochemistry (that is, in the shapes that their adducts assume); and in oxidation state (that is, in the charge that they carry). Nature has chosen transition metals to serve as the centerpieces for many of its catalysts, the enzymes, most probably for these same reasons.

COMPLEX REACTIONS

Parallel Processes, Consecutive Processes, and Reversible Processes. Thus far we have discussed relatively simple reactions, which occur by mechanisms that involve one step that is substantially slower than the others, and which go to completion. In this situation, rate laws may be readily written from the mechanism and compared with the laboratory-determined rate law. As you might expect, most reactions are not so simple. In this section, we briefly describe three frequently encountered complications: parallel reactions (two or more simultaneous processes leading to the same or different products); consecutive reactions having comparable rates; reversible reactions (those in which the reverse process plays an important role in determining the observed rate).

PARALLEL REACTIONS

Parallel reaction paths may be represented schematically as in 15–8–1.

$$A + B \rightarrow P\ [k_1]$$
$$A + C \rightarrow Q\ [k_2]$$

To simplify discussion, we assume that each reaction occurs by a single bimolecular elementary step. If so, then we may write the rate laws directly. The rate of disappearance of A is the sum of the rates at which it is used in 15–8–1a and 15–8–1b:

$$\text{Rate} = -d[A]/dt = k_1[A][B] + k_2[A][C] = \{k_1[B] + k_2[C]\}[A]$$

If pseudo-order conditions are imposed, this simplifies to the pseudo first-order rate law in 15-8-3.

$$\text{Rate} = -d[A]/dt = (k_1' + k_2')[A]$$

where $k_1' = k_1[B]$ and $k_2' = k_2[C]$. A plot of Rate versus [A] is linear, with slope $k_1' + k_2'$. Thus if one follows the reaction rate by monitoring [A], only the sum of the two rate constants is obtained. If it is possible to monitor [P] or [Q], k_1' or k_2' may be separately determined, and the other one calculated by difference.

Example. The overall reaction below is thought to occur by two different pathways operating simultaneously.

$$Pt(NH_3)_2Cl_2 + H_2O \rightarrow Pt(NH_3)_2Cl(H_2O)^+ + Cl^-$$

The first pathway involves attack by H_2O at the platinum atom, with simultaneous loss of Cl–. Product is thus formed in a single elementary step with the same equation as the overall reaction, with rate constant k_1. In the second pathway, Cl^- leaves the platinum compound in a slow first step; this is followed by a rapid second step in which H_2O forms a bond with platinum:

$$Pt(NH_3)Cl_2 \rightarrow Pt(NH_3)Cl^+ + Cl^-\ [k_1,\ \text{slow}]$$
$$Pt(NH_3)_2Cl^+ + H_2O \rightarrow Pt(NH_3)_2Cl(H_2O)^+\ [k_2,\ \text{rapid}]$$

What rate law will be exhibited if this hypothesis is correct?

Solution. The overall rate law is the sum of the rate laws for the two parallel processes:

$$\text{Rate} = \text{Rate mechanism 1} + \text{Rate mechanism 2}$$
$$= k_1[Pt(NH_3)_2Cl_2][H_2O] + k_2[Pt(NH_3)_2Cl_2]$$

(Be sure that you understand why the rate laws for the two mechanisms have these forms.)

CONSECUTIVE REACTIONS

We have previously examined systems consisting of two consecutive reactions under what might be called limiting conditions: either the first reaction was much more rapid than the second and could be treated as an equilibrium; or the first reaction was much slower than the second and completely determined the overall reaction rate. It is instructive to examine how the concentrations of A, B, and C vary with time in these situations, as well as the one in which the two steps have comparable rate constants.

Consider the following two step mechanism:

$$A \rightarrow B\ [k_1]$$
$$B \rightarrow C\ [k_2]$$

We present three situations involving different relative sizes of k_1 and k_2. We assume for simplicity that only A is present initially; [B] = [C] = 0 at t = 0. When $k_1 >> k_2$, reaction goes to completion before 15-8-4b gets under way. When $k_1 = 20k_2$, the concentrations of A, B, and C vary as shown in Figure. A rapidly disappears and is replaced by B; B then very slowly converts to C.

If the reaction rate is monitored by measuring the time dependence of [C], only the rate constant k_2 is obtained. On the other hand, if [A] is monitored, its rapid disappearance is governed by k_1, which is obtainable from either a plot of ln [A] versus time, or a plot of -d[A]/dt vs [A]. When $k_1 << k_2$, B is formed very slowly via 15–8–4a, but is then immediately consumed by 15–8–4b.

The concentration of B thus never rises much above zero, and C appears at the same rate that A disappears. This corresponds to the second set of limiting conditions mentioned above. When $k_2 = 20k_1$, the concentrations of the 3 species vary as shown in Figure. The rate constant determined experimentally is k_1, the smaller of the two. In this situation, it is very difficult to measure k_2 as [B] is never appreciable.

Only if B can be prepared independently under conditions where it does not convert to C can the rate constant, k_2, be measured.

There are many situations in which k_1 and k_2 are approximately equal. We illustrate these by looking at the case in which $k_1 = k_2$. The time dependence of the concentrations of A, B, and C is presented in Figure. At the very beginning of reaction, t = 0, the rate of the second reaction is zero because [B] = 0. The initial rate is the rate of disappearance of A, from which the rate constant k_1 can be determined.

Before reaction has progressed very far, however, [B] has risen sufficiently that the second process occurs simultaneously with the first. This tends to drain the concentration of B as it is converted to C. In the Figure, we see [B] rise to a maximum, then fall off because its rate of production from A slows as A is consumed. At any time, it must be true that

[A] + [B] + [C] = constant = $[A]_o$. If [C] is monitored and [A] is calculated from the value of k_1, $[A]_o$, and t, [B] can be calculated as a function of time. It is then possible to determine k_2 from appropriate plots.

Finally, the mechanism in 15-8-4a and b can be modified slightly to allow reversibility in the first (but not the second) step:

$$A \Leftrightarrow B\ [k_1, k_{-1}]$$
$$B \rightarrow C\ [k_2]$$

(When k_{-1} = 0 this mechanism reduces to the one in 15–8–4a and b). Figure 10.20 shows the time dependence of the concentrations of A, B, and C assuming k_1, $k_{-1} >> k_2$.

This corresponds to the first limiting situation mentioned at the beginning of this section. Figure shows the concentration behaviour for a short time early in the reaction. It is very clear from this figure that A and B rapidly come to equilibrium and maintain a constant concentration ratio thereafter. The total of their concentrations is then very slowly reduced as B converts to C. Figure shows the course of reaction over a longer time.

The plots in Figures 10.18 to 10.20 were obtained using finite difference methods to calculate the concentrations of the three species, A, B, and C, at many times during the course of reaction.

These repetitive calculations were carried out using an electronic spreadsheet.

Table–2 contains the data used in plotting outlines the method of calculation programmed into the spreadsheet. It would be instructive for you to set up your own spreadsheet according to the information in to see that you can produce plots similar to those in Figures 10.18 to 10–20.

REVERSIBLE REACTIONS

Our last example of complexity in chemical dynamics will focus on reversible reactions: those that approach an equilibrium position in which appreciable amounts of both reactants and products are present. The simplest such process is shown in 15/08–6.

$$A \Leftrightarrow B\ [k_1, k_{-1}]$$

k_1 and k_{-1} are the rate constants for the forward and reverse processes, respectively, which are both assumed to be first order. The rate law in 15–8–7 follows:

$$\text{Rate} = -d[A]/dt = k_1[A]-k_{-1}[B]$$

This equation states that the rate of change in the concentration of A is the difference between its rate of consumption via the forward process and its rate of production via the reverse process.

If the initial concentration of B is 0, then at any point during the process, [B] = $[A]_o$–[A].Equation 15-8-7 becomes

$$-d[A]/dt = (k_1 + k_{-1})[A]-k_{-1}[A]_o$$

This can be integrated, but the resulting expression is not readily plottable. Thus, even though the reaction is first-order, a plot of ln[A] versus t will not be linear. To eliminate the troublesome $k_{-1}[A]_o$ term, we recognize that 15–8–8 applies at all points during reaction, including equilibrium. Letting $[A]_e$ represent the equilibrium concentration of A, 15–8–9 holds when the reaction reaches equilibrium:

$$-d[A]_e/dt = (k_1 + k_{-1})[A]_e - k_{-1}[A]_o$$

Subtracting 15–8–9 from 15-8-8 eliminates the $k_{-1}[A]_o$ term, giving,

$$-d([A]-[A]_e)/dt = (k_1 + k_{-1})([A]-[A]_e) = k_{tot}([A]-[A]_e)$$

This equation is readily integrated in the variable $[A]-[A]_e$, which is the *deviation* of the concentration of A from the equilibrium value. Thus 15–8–10 says that in a reversible first-order process, the approach to equilibrium is governed by first-order kinetics. A plot of ln $([A]-[A]_e)$ against t will be linear, with slope $-k_{tot}$, the sum of the forward and reverse rate constants.

We make several observations about reversible first-order processes:

- The rate constant for approach to equilibrium is the sum of forward and reverse rate constants because equilibrium is approached by simultaneous reaction in both directions.
- When 15-8-6 is irreversible, $k_{-1} = 0$ and $[A]_e = 0$; thus 15-8-10 reduces to 15–3–6 (with n = 1).
- When $k_1 \gg k_{-1}$, the measured rate constant, k_{tot}, is essentially equal to k_1. The reverse process is unimportant until the reaction is nearly at equilibrium, because $k_{-1}[B]$ remains small. At equilibrium, the forward and reverse reactions, given by the following expressions, are equal:

$$Rate_f = k_1[A] = Rate_r = k_{-1}[B]$$

Both rates are small since at equilibrium [A] and k_{-1} are both small. Similar considerations hold when $k_1 \ll k_{-1}$. In this case, the measured rate constant is essentially k_{-1}.

Let's examine a particular example of a reversible first order process.

Example: The kinetics of the reaction below were monitored by following the change in absorbance with time of R.

$$R \Leftrightarrow P$$

The following data were obtained.

Time, Min	**Absorbance Due to R**
1	.789
2	.648
4	.452
6	.333
8	.261
10	.217

12	.191
60	.150

Solution. Determine the rate constant for approach to equilibrium. The absorbance value at 60 minutes represents the equilibrium value, since enough time has elapsed for the reaction to have reached equilibrium. Thus $A_e = 0.15$. A value of $A\text{-}A_e$ is calculated for each time and a plot of $\ln(A–A_e)$ constructed. Figureshows such a plot. The linearity of the plot substantiates a first order approach to equilibrium. The slope of the plot gives the rate constant, 0.27 min^{-1}. This is actually the sum of the forward and reverse rate constants for the reaction. More complicated processes can be made to conform to 15–8–10 by imposing pseudo-order conditions. For example, reaction 15–8–11 involves forward and reverse processes that are second order:

$$A + B \Leftrightarrow C + D\ [k_1, k_{-1}]$$

Suppose that initially $[B]_o$ and $[D]_o$ are both made much larger than $[A]_o$, and $[C]_o = 0$. Then the concentrations of B and D will remain constant during the reaction, and A will be converted to C in a pseudo-first-order process in which the forward rate constant, k_f, is equal to $k_1[B]$; and the reverse rate constant, k_r, is equal to $k_{-1}[D]$. By analogy with 15-8-10, we write the rate law for approach to equilibrium:

$$\text{Rate} = –d([A]–[A]_e)/dt = (k_f + k_r)([A]–[A]_e) = k_{tot}([A]\ [A]_e)$$

k_{tot} is studied first as a function of [B] with [D] held constant, in which case 15-8-13 applies:

$$k_{tot} = k_1[B] + k_r$$

k_1 can be obtained from a plot of k_{tot} versus [B]. Then k_{tot} is studied as a function of [D] with [B] held constant, when 15–8–14 applies:

$$k_{tot} = k_f + k_{-1}[D]$$

k_{-1} is obtained from a plot of k_{tot} versus [D].

ENZYMES AS CATALYSTS

Even the simplest organism is powered by thousands of chemical reactions, participating in a complex, interconnected web of chemical activity.

On their own, most of these reactions occur with rates so slow as to seem inconsistent with life. In the organism, however, they take place briskly and *specifically* (*i.e.*, only the desired products are formed), thanks to nature's catalytic molecules, the enzymes.

Enzymes are a subclass of biopolymers called proteins. They consist of very long molecules called polypeptides, which are formed by stringing amino acids together. In general they are huge molecules, with molecular masses in the range between 14000 and 8 * 10^6 μ. Needless to say, the structures and stereochemistries of these macromolecules are very complex. The enzyme, chymotrypsin, which is small as enzymes go, is shown in Figure. In the figure, gray balls are carbon, red are oxygen, blue are nitrogen, and yellow are sulfur atoms; hydrogen atoms are not shown. It is estimated that the human body contains more than 2000 enzymes, each with its own unique structure, each catalyzing only a single reaction or closely related set of reactions. Enzymes are extremely selective—specific—in their catalysis.

We will have more to say about enzyme structure in a subsequent chapter. In this section, our focus will be on the kinetics of enzyme catalysis. Despite the vast number of enzyme structures and the equally vast number of reactions catalyzed, most enzyme-catalyzed processes exhibit essentially the same kinetics. Considering the complexity of the molecules and reactions, the kinetics is relatively simple. We approach it first from the experimental standpoint, which produces an observed rate law; we then address the question of mechanism. Equation shows a typical enzyme catalyzed reaction in generic form.

$$\text{Substrate} \xrightarrow{E} \text{Products}$$

Here, E stands for the enzyme, and is written over the arrow to indicate that the reaction is catalyzed. The reactant molecule on which the enzyme works is called the *substrate*. Although there may be more than one reactant molecule, in most cases only one of them interacts directly with the enzyme. We will condense 15–9–1 to the form in 15–9–2.

$$S \xrightarrow{E} P$$

Kinetics studies of enzyme catalyzed reactions usually involve the measurement of initial rates, for reasons that need not concern us here. Throughout the discussion that follows, it will be assumed that the word, rate, stands for initial rate. The general rate law, applicable to most enzyme catalyzed processes for which studies have been done, is in 15–9–3.

$$\text{Rate} = k_{obs}[E]_T$$

Here $[E]_T$ stands for the total concentration of enzyme present in the solution, and k_{obs} is the observed first-order rate constant. The observed rate constant exhibits an interesting dependence on the concentration of substrate, S. When [S] is small, k_{obs} increases nearly linearly with [S]; in other words, the rate is first order in S. However, as larger and larger concentrations of S are used, the increases in k_{obs} become smaller and smaller; k_{obs} eventually levels off to a constant value that is independent of further increases in [S].

At this point, the rate is zero-order in S. The relationship between k_{obs} and [S] is shown in Figure. To an experienced kineticist, this behaviour is known as *saturation kinetics*, and can be explained by a rate law of the form in 15–9–4.

$$\text{Rate} = k_{obs}[E]_T = A[E]_T[S]/(B + [S])$$

Here A and B are constants that are presumably related to rate constants of the elementary steps in the reaction mechanism. We interpret the observed kinetics in terms of 15-9-4 as follows. When [S] is small, B is the dominant term in the denominator. Under these conditions, 15–9–4 simplifies to

$$\text{Rate} = A[E]_T[S]/B$$

Thus the rate is first order in E, first order in S, with k_{obs} = A/B. When [S] is large, it is the dominant term in the denominator, and 15–9–4 simplifies to

$$\text{Rate} = A[E]_T$$

The rate is first order in E, zero-order in S, with k_{obs} = A. The form of the expression in is consistent with the observed kinetics. For historical reasons, enzyme-substrate systems that obey the kinetic form in 15–9–4 are said to exhibit *Michaelis-Menten kinetics*, after two early investigators in the field. The constant, B, is called the Michaelis constant, and is usually given the symbol, K_M. We will not use this symbol, however, because it implies that B is an equilibrium constant. Instead, we will use the symbol M. We make the switch at this point:

$$\text{Rate} = A[E]_T[S]/(M + [S])$$

$$k_{obs} = A[S]/(M + [S])$$

The plot of Rate versus substrate concentration nicely shows the saturation behaviour of the kinetics; unfortunately, it is not possible to obtain accurate values for A and M from this plot. However, there are two rearrangements of 15–9–8 that allow the experimental data to be plotted in linear form. The first of these is obtained by reciprocating both sides of 15–9–8:

$$1/k_{obs} = M/A[S] + 1/A$$

A plot of $1/k_{obs}$ versus 1/[S] should be linear with slope M/A and intercept 1/A. Figure shows such a plot for the data of Figure. A reciprocal plot of this type is called a Lineweaver-Burk plot. The values of A and M obtained from are A = 0.01 s^{-1}, M = 0.0055 M. The second linear form is obtained from 15–9–8 by, first, multiplication of both sides by k_{obs};

$$(15–9–10): 1 = k_{obs}M/A[S] + k_{obs}/A$$

second, multiplication by A;

$$(15–9–11): A = k_{obs}M/[S] + k_{obs}$$

and third, subtraction of the first term on the right from both sides.

$$(15–9–12): k_{obs} = A – k_{obs}M/[S]$$

A plot of k_{obs} versus k_{obs}/[S] should be linear with slope -M and intercept A. This plot, called an Eadie-Hofstee Plot, is shown in. It produces M = 0.0055 M, A = 0.01 s^{-1}.

The physical interpretation of Michaelis-Menten (saturation) kinetics goes something as follows. First, the rate is first order in both enzyme and substrate when [S] is small. This implies that 1 molecule of enzyme and 1 molecule of substrate must collide and interact in order to form products. Second, the rate is first order in E and zero-order in S when S is large.

The implication of this observation is that the substrate forms an adduct of some type with the enzyme, which we might symbolize as ES. This adduct then reacts further to form products. When [S] becomes large enough, Le Chatelier's Principle requires that all of the free enzyme is converted to the adduct. Once this happens, the rate no longer increases with increases in [S], because [ES] has reached its maximum possible value. This type of reasoning led biochemists to the simplest mechanism consistent with the observations:

$$E + S = ES\ [k_1, k_{-1}]$$

$$ES \rightarrow P + E\ [k_{cat}]$$

We must now develop the rate law implied by this mechanism, to see that it matches what is observed. It is not legitimate to make assumptions at the outset about the relative magnitudes of the rate constants, because we have no information about that. The development of the rate law must be carried out as generally as possible, then examined for limiting behaviour. We use the so–called *steady state approximation*. The essence of this approximation is that *the concentration of the intermediate adduct, ES, stays essentially constant as reaction proceeds*. This idea is fairly simple to implement. First, we recognize that the rate of formation of products, P, is given by,

$$\text{Rate} = d[P]/dt = k_{cat}[ES]$$

This rate law is strictly correct and always true within the context of the mechanism in 15–9–13. However, it is difficult to use because [ES] is not accessible. As usual, we need to express the rate law in terms of measurable concentrations: those of E and S. Further work is required. Now we invoke the steady state assumption, which is that,

$$d[ES]/dt = 0$$

This is a mathematical way to say what we said above: the concentration of ES remains constant as reaction proceeds. We now use the elementary steps to obtain an expression for [ES]:

$$d[ES]/dt = 0 = k_1[E][S] - k_{-1}[ES] - k_{cat}[ES]$$

Solving for [ES] gives,

$$[ES] = k_1[E][S]/(k_{cat} + k_{-1})$$

which can be substituted for [ES] in the rate law, 15-9-14.

$$\text{Rate} = k_1 k_{cat}[E][S]/(k_{cat} + k_{-1})$$

We are close, but not quite done. Recall that the experimental rate law is written in terms of the total enzyme concentration, $[E]_T$. Interpreted in terms of our mechanism,

$$[E]_T = [E] + [ES]$$

In words, the enzyme is partitioned between the unbound and substrate bound forms. We cannot at the outset expect to know how much enzyme is in each form. However, we do know the total amount; it is the amount we put into the solution. We must therefore obtain an expression for [E] in terms of $[E]_T$, which can be done using 15-9-17 and 15–9–19:

$$[E]_T = [E] + [ES] = [E] + k_1[E][S]/(k_{cat} + k_{-1})$$

Solving for [E] gives,

$$[E] = [E]_T/\{1 + (k_1[S]/(k_{cat} + k_{-1})\}$$

Substituting this expression into the rate law, 15-9-18, and doing some simplifying arithmetic gives

$$\text{Rate} = k_{cat}[E]_T[S]/\{[S] + (k_{cat} + k_{-1})/k_1\}$$

This has a familiar look. In fact, if we identify k_{cat} with A, and $(k_{cat} + k_{-1})/k_1$ with M, we obtain

$$\text{Rate} = A[E]_T[S]/\{M + [S]\}$$

which is identical in form with the experimental law.

Within the context of mechanism, we can draw the following conclusions:

- The value of A gives a direct measure of the rate constant k_{cat}, which governs the rate at which the enzyme substrate adduct reacts to form products.
- The Michaelis constant, M, is seen to be $(k_{cat} + k_{-1})/k_1$. This gives the ratio of the rate of consumption of ES to its rate of production. Importantly, it is clear that M is not an equilibrium constant, because it involves the sum of rate constants from two different elementary processes.

MECHANISMS OF SUBSTITUTION REACTIONS

A substitution reaction is represented generally in equation 1:

$$(15\text{–}10\text{–}1)\text{: A-X + Y} \rightarrow \text{A-Y + X}$$

This reaction type, in which a group Y replaces a group X in a molecule, is widespread in chemistry. Several examples of substitution reactions are presented in equations 2-5.

$$(15\text{–}10\text{–}2)\text{: H–CN} + F^- \rightarrow HF + CN^-$$

$$(15\text{–}10\text{–}3)\text{: } CH_3CH_2Cl + OH^- \rightarrow CH_3CH_2OH + Cl^-$$

$$(15\text{–}10\text{–}4)\text{: } (NH_3)_5CoCl^{2+} + H_2O \rightarrow (NH_3)_5CoOH_2{}^{3+} + Cl^-$$

$$(15\text{–}10\text{–}5)\text{: } PCl_5 + H_2O \rightarrow P(OH)_5 + HCl$$

Reaction (2) is a Bronsted-Lowry acid base reaction, discussed. It is also a substitution reaction in which one base, F^-, replaces another base, CN^-, in forming a bond to the proton, H+. Reaction (15–10–3) shows the formation of an alcohol from an alkyl halide by substitution of a hydroxide anion for a chloride anion at carbon. Reaction (15–10–4) involves the replacement of one *ligand* by another in a transition metal complex. We will discuss these in more detail below. Finally, (15–10–5) is a double substitution (or double displacement) process in which OH^- and Cl^- substitute for each other. Because substitution reactions are so common, they have attracted the attention of chemists for many years. Of particular interest has been the mechanism(s) by which they take place. What are the details of the process the net result of which is replacement of X by Y? Do all substitution reactions occur in the same way, or are there different pathways that depend in some understandable way on the structures of A, X, and Y?

In this section we will explore these questions and attempt to supply some answers. As we will see, kinetics studies have played the central role in the discovery of mechanism. Our discussion will parallel the historical development of the subject, beginning with the mechanisms of substitution reactions at carbon.

SUBSTITUTION REACTIONS AT CARBON

The replacement of a halide by a hydroxide anion is shown in 15-10-6:

$$(15–10–6): R–Cl + OH^- \rightarrow ROH + Cl^-$$

R represents a carbon-containing group of atoms that may be simple (for example, CH_3) or more complex (for example, $(CH_3)_3C^-$). (15–10–3) is a specific example of the general process in (15–10–6). It has been shown via many research studies over many years that (15–10–6) occurs by two different pathways, depending on the nature of the carbon-containing group R.

The Second-Order Pathway

Consider first the kinetics of (15-10-3), in which R can be identified with the CH_3CH_2– portion of the reactant, CH_3CH_2Cl. The reaction can be studied under pseudo-order conditions in CH_3CH_2Cl. Recall that this means that the concentration of OH- is maintained at a value at least 10 times larger than the concentration of CH_3CH_2Cl, so that $[OH^-]$ changes imperceptibly during reaction. Studied in this manner, the kinetics are first-order in CH_3CH_2Cl, with a pseudo-first order rate constant, k', that is directly proportional to the concentration of OH^-. The experimental rate law thus has the form in (15-10-7).

$$\text{Rate} = k_{obs}[CH_3CH_2Cl] \text{ where } k_{obs} = k[OH^-]$$

Thus (15–10–3) is experimentally observed to follow second-order kinetics. From this experimental basis, the chemist must then speculate as to the possible mechanisms that could give rise to a rate law of this form.

In other words, the chemist must propose a theory about the elementary steps involved in the reaction process. Reaction seems quite simple, so it is conceivable that it could occur in a single step in which the Cl^- ion leaves simultaneously with attack by the OH^- ion. In other words, (15–10–3) could be an elementary process. As we know, such processes have rate laws that follow directly from their stoichiometries. Thus, IF were to occur in the single step (15-10-7), it would necessarily obey the rate law.

$$CH_3CH_2Cl + OH^- \rightarrow HO—CH_3CH_2—Cl \rightarrow CH_3CH_2OH + Cl^- [k_1] \rightarrow$$
$$Rate = -d[RCl]/dt = k_1 [RCl] [OH^-]$$

This agrees with the experimental rate law, and so the single-step mechanism is consistent with experiment. However, the thorough chemist would not stop here. Rather, s/he would ask whether there are other mechanisms that are consistent with the experimentally observed second-order kinetics. In fact, there is at least one other mechanism that should be considered. It is the two-step mechanism:

$$CH_3CH_2Cl + OH^- \Leftrightarrow CH_3CH_2(OH)Cl^- [k_1, k_{-1}]$$
$$CH_3CH_2(OH)Cl^- \rightarrow CH_3CH_2OH + Cl^- [k_2]$$

In the first step of this process, the OH– group forms a bond to the carbon atom that is attached to Cl^-, forming an intermediate in which this carbon atom is attached to 5 other groups.

In the second step, the bond to the Cl^- ion breaks, generating the products. The reverse arrow in the first step is included to recognize that the just-formed bond to the OH– ion could also break, regenerating the reactants. A steady–state treatment of the mechanism, in which the concentration of the intermediate is assumed to stay constant during the course of reaction, yields the theoretical rate law in (15–10–10):

$$Rate = k_1k_2[CH_3CH_2Cl][OH–]/(k_{-1} + k_2)$$

The 2-step mechanism in (15–10–9) also predicts second-order kinetics. In this case, the experimental k_{obs} would be identified with the expression $k_1k_2/(k_{-1}+k_2)$, which involves the rate constants of all of the elementary steps in mechanism. The important point here is that the experimental rate law does not enable us to choose between the one-step elementary mechanism and the more elaborate two-step mechanism.

Indeed, there are probably other even more elaborate mechanisms that also predict second-order kinetics. Any Mechanism That Predicts A Rate Law Having The Same Form As The Experimental Rate Law Must Be Considered As Possible. Experiments of other types must be designed in an attempt to narrow the field of possible mechanisms. Many such experiments have been designed and carried out. Based on the results, most chemists now believe that reaction (15–10–3) occurs by the elementary mechanism.

Because reaction is only a specific example of reactions of type, is it then reasonable to conclude that all reactions of type (15–10–6) occur by the same single-step process? As is so often the case in science, the situation is not nearly so clean and simple! Thus, although many substitution reactions at carbon show second-order kinetics and are therefore consistent with the one-step mechanism, other reactions show rate laws that are first order in RCl and zero-order in OH–. An example.

$$(CH_3)_3CCl + OH^- \rightarrow (CH_3)_3COH + Cl^-$$

Kinetics studies of (15-10-11) yield the experimental rate law

$$\text{Rate} = k_{obs}[(CH_3)_3CCl]$$

Interestingly, the rate of (15–10–11) is independent of how much OH^- is made availa and that the slow step does not involve the incoming group OH^-proposed to explain this behaviour?

First-order kinetics implies that the process occurs in at least two steps, and that the slow step does not involve the incoming group OH^-. One mechanism that produces a rate law of the form of (15-10-12) is given in (15-10-13).

$$(CH_3)_3CCl \rightarrow (CH_3)_3C^+ + Cl^- \; [k_1 \text{ (slow)}]$$

$$(CH_3)_3C^+ + OH^- \rightarrow (CH_3)_3COH \; [k_2]$$

Because the overall rate is determined by the rate of the slowest (rate-determining) step, the theoretical rate law for mechanism (15–10–13) is

$$\text{Rate} = k_1[(CH_3)_3CCl]$$

This is consistent with the experimental rate law.

Mechanism is oversimplified, however, because it ignores the possibilty that the Cl– ion could reattach to the carbon species, $(CH_3)_3C^+$. We can fix this by including the reverse process. The amended mechanism is in (15-10-15).

$$(CH_3)_3CCl \Leftrightarrow (CH_3)_3C^+ + Cl^- \; [k_1, k_{-1}]$$

$$(CH_3)_3C^+ + OH^- \rightarrow (CH_3)_3COH \; [k_2]$$

A steady-state treatment on the intermediate, $(CH_3)_3C^+$, gives the rate law in (15–10–16).

$$\text{Rate} = k_1k_2[(CH_3)_3Cl][OH^-]/(k_{-1}[Cl^-] + k_2[OH^-])$$

Allowing for the reverse of the first step therefore greatly complicates the theoretical rate law. The most noticeable feature of (15–10–16) is that it is no longer obvious that the kinetics are simple first-order in $(CH_3)_3CCl$. Instead, the overall reaction order depends on the relative size of the 2 terms in the denominator.

Under conditions where $k_2[OH^-] \gg k_{-1}[Cl^-]$, the $k_2[OH^-]$ term dominates and will cancel with the $k_2[OH^-]$ term in the numerator. The resulting rate law is in:

$$\text{Rate} = k_1[(CH_3)_3Cl]$$

This is consistent with the observed first order behaviour. However, under conditons where $k_2[OH^-] \ll k_{-1}[Cl^-]$, the latter term will dominate the denominator, and the rate law becomes

$$\text{Rate} = k_1k_2[(CH_3)_3CCl][OH^-]/k_{-1}[Cl^-]$$

Under conditions of small [OH–], the rate should be first order in $(CH_3)_3Cl$, first order in OH^-, and inverse first order in Cl^-. Indeed, this suggests that this mechanism could be tested by studying the kinetics using a small $[OH^-]$. Observation of an inverse dependence on $[Cl^-]$ would provide support for mechanism.

In summary, to account for the varied kinetics behaviour displayed by reactions of type (15–10–6), chemists have proposed two limiting mechanisms. These are shown below with their traditional names:

The S_N2 Mechanism:

$$\text{R–X} + \text{Y} \rightarrow (\text{Y—R—X}) \rightarrow \text{R–Y} + \text{X}$$

The S_N1 Mechanism:

$$\text{R–X} \rightarrow \text{R} + \text{X}$$

$$\text{R} + \text{Y} \rightarrow \text{R–Y}$$

The abbreviation, S_N2, stands for *substitution nucleophilic second-order*. This name is a mouthful, but it is easily explained. “Substitution” and “second-order” are self-explanatory.

The adjective nucleophilic (nucleus-loving) is used to describe an atom, group of atoms, or ion that is attracted to a positive or partially-positive center in a molecule. Both X and Y are nucleophilic groups that compete for the positive center, R+.

In the S_N2 mechanism, Y attacks the carbon atom in R–X and forms a weak bond. Simultaneously, the bond between R and X weakens. The result is a transition state, Y—R—X, in which both the leaving group (X) and the entering group (Y) are weakly bonded to R. The reaction is completed when the bond to X is severed and replaced with a full bond to Y.

The reaction thus occurs in a single elementary step and therefore must display second-order kinetics, consistent with its stoichiometry. The abbreviation, S_N1, stands for *substitution nucleophilic first-order*. The name suggests a substitution involving the competition of 2 nucleophiles for a positive center that displays first order kinetics.

SUBSTITUTION IN TRANSITION METAL COMPLEXES

Elements in the d block of the periodic table, sandwiched between the s block elements at the left and the p-block elements at the right, are called transition metals. These elements form a class of compounds in which the transition metal is bonded to a number of ions or molecules via covalent bonds. When attached to a transition metal ion, such ions or molecules are called ligands. Most ligands are also Bronsted-Lowry bases, so that we can recognize them by the availability of at least one lone pair of electrons. In fact, it is a lone pair of electrons on the ligand that is used to form the bond to the transition metal ion. Some examples of transition metal complexes are given below.

$Fe(CN)_5(NO)^{3-}$. The transition metal ion is Fe^{2+}. It is attached to 6 ligands, five of them cyanide anions (CN^-) and the sixth a nitric oxide molecule. The complex has a net charge of 3–, obtained by summing the charges on the metal ion and the ligands.

The six ligands are arranged about the metal ion in an octahedron. The complex is brown in colour.

$Co(NH_3)_5Cl^{2+}$. The transition metal ion is Co^{3+}. It is attached to 6 ligands, five of them ammonia molecules (NH_3) and the sixth a chloride anion. The net charge of 2+ comes from summing the metal charge (3+) and the chloride charge (1-). The six ligands are arranged about the metal ion in an octahedron. The complex is green.

$CoCl_4^{2-}$. The metal ion is Co^{2+}, which is attached to four Cl^- ligands arranged about it in a tetrahedron. The overall charge is the sum of the ion charge (2+) and the chloride charges (4–). The complex is intensely blue in colour.

$Ni(CN)_4^{2-}$. The metal ion is Ni^{2+}, attached to four cyanide ligands which form a square planar arrangement. The complex is yellow in colour.

These few examples illustrate some generalities about transition metal complexes:

- Many of them are coloured.
- Common numbers of ligands are 6 and 4.
- Six ligands almost invariably form an octahedron. Four ligands form either a tetrahedron or a square plane.

The great variety of colors exhibited by transition metal complexes has captured the interest of chemists for centuries. Long ago it was observed that colour changes sometimes occur when a complex is put in the presence of another potential ligand different from those already contained in it. The colour changes are the result of substitution of one ligand by another, as for example in equation (15-10-19):

$$CoCl_4^{2-}(\text{blue}) + 6H_2O \rightarrow Co(H_2O)_6^{2+}(\text{pink}) + 4\ Cl^-$$

The mechanisms of ligand substitution reactions have been intensely investigated since the 1950s. The rate at which one ligand substitutes for another depends tremendously on what the metal is, and on what the other ligands are.

In complexes of Cu^{2+}, one ligand can substitute for another at a rate of up to 10^{10} times per second, whereas in complexes of Co^{3+} and Cr^{3+}, substitution can take hours or days. It is not surprising that study of complexes of the latter two ions has been very popular, and quite a bit is now known about their mechanisms of substitution.

Substitution at Co^{3+}. Consider reaction (15–10–20), in which a water ligand is replaced by a chloride ion in aqueous solution:

$$Co(NH_3)_5(H_2O)^{3+} + Cl^- \rightarrow Co(NH_3)_5Cl^{2+} + H_2O$$

Kinetics studies of this reaction give the rate law, Rate = $k_{obs}[Co(NH_3)_5(H_2O)][Cl^-]$. This is consistent with several mechanisms, shown below:

Mechanism 1, one-step:

$$Co(NH_3)_5(H_2O)^{3+} + Cl^- \rightarrow Co(NH_3)_5(H_2O)(Cl)^{2+} \rightarrow Co(NH_3)_5Cl^{2+} + H_2O$$

This mechanism implies the rate law, Rate = $k_1[Co(NH_3)_5(H_2O)^{3+}][Cl^-]$, where k_1 is the rate constant for the single elementary step process shown.

Mechanism 2, two-step:

$$Co(NH_3)_5(H_2O)^{3+} \Leftrightarrow Co(NH_3)_5{}^{3+} + H_2O\ [k_1\ (\text{slow}),\ k_{-1}\ (\text{fast})]$$

$$Co(NH_3)_5{}^{3+} + Cl- \rightarrow Co(NH_3)_5Cl^{2+}\ [k_2,\ \text{fast}]$$

The steady-state assumption for the intermediate, $Co(NH_3)_5{}^{3+}$, yields the rate law in (15-10-23).

$$\text{Rate} = k_1k_2[Co(NH_3)_5(H_2O)^{3+}][Cl-]/(k_{-1}[H_2O] + k_2[Cl-])$$

In aqueous solution, $[H_2O]$ is huge and constant, so the first denominator term is expected to be much larger than the second term. The rate law then simplifies to

$$\text{Rate} = k_1k_2[Co][Cl-]/k_{-1}[H_2O] = k_{theor}[Co][Cl-]$$

which agrees with the observed rate law if $k_{obs} = k_1k_2/k_{-1}[H_2O]$. This is called *the Dissociative Mechanism*, because the outgoing ligand first dissociates from the metal before the incoming ligand bonds.

Mechanism 3, two-step:

$$Co(NH_3)_5(H_2O)^{3+} + Cl^- \Leftrightarrow Co(NH_3)_5(H_2O)(Cl)^{2+}\ [k_1, k_1]$$

$$Co(NH_3)_5(H_2O)(Cl)^{2+} \rightarrow Co(NH_3)_5Cl^{2+} + H_2O\ [k_2]$$

The implied rate law is,

$$\text{Rate} = k_1k_2[Co(NH_3)_5(H_2O)^{3+}][Cl^-]/(k_{-1} + k_2)$$

which agrees with the observed rate law if $k_{obs} = k_1k_2/(k_{-1}+k_2)$. This is called the *Associative Mechanism*, because the incoming ligand is proposed to first associate with the metal before the outgoing ligand leaves.

Mechanism 4, two-step:

$$Co(NH_3)_5(H_2O)^{3+} + Cl^- \Leftrightarrow [Co(NH_3)_5(H_2O)\cdots Cl]^{2+} \; [K_{eq}]$$
$$[Co(NH_3)_5(H_2O)\cdots Cl]^{2+} \rightarrow Co(NH_3)_5Cl^{2+} + H_2O \; [k_2, \text{slow}]$$

The implied rate law for this mechanism is,

$$\text{Rate} = k_2K_{eq}[Co]_T[Cl^-]/(1 + K_{eq}[Cl^-])$$

where $[Co]_T$ represents the sum of the concentrations of $Co(NH_3)_5(H_2O)^{3+}$ and $[Co(NH_3)_5(H_2O]\cdots Cl]^{2+}$. This, too, predicts second order kinetics under conditions where $K_{eq}[Cl^-] \ll 1$. This mechanism is called the *Interchange mechanism*, because the RDS involves an interchange of the incoming and outgoing ligands at the metal.

How are these mechanisms to be distinguished? Examination of the implied rate laws provides the answer. Both the dissociative and the interchange mechanisms predict that at sufficiently high concentration of the incoming ligand, the rate should become zero-order in incoming ligand. This is because the denominator term involving the incoming ligand should eventually become large enough to dominate, and would then cancel the incoming ligand factor in the numerator.

This behaviour has been experimentally observed in the reaction of a few complexes of Co^{3+} with anions, and on this basis the one-step and associative mechanisms have been tentatively ruled out for ligand substitution at this metal ion. It is now accepted by most chemists that substitution in Co^{3+} complexes occurs by the interchange mechanism because, despite much work, an intermediate of the type proposed in the dissociative mechanism has not been experimentally observed. Substitution at Fe^{2+}. In some ligand environments, substitution reactions at the Fe^{2+} ion are slow enough to be conveniently measured at room temperature. The complex, $Fe(CN)_5X^{3-}$, where X is a variety of neutral molecules, is particularly convenient to study because the CN^- ligands are very difficult to replace. Reactions in which X is replaced by some other group, 0

$$Fe(CN)_5X^{3-} + Y \rightarrow Fe(CN)_5Y^{3-} + X$$

It has been shown experimentally that the reaction obeys a rate law of the form in (15–10–30):

$$\text{Rate} = k_{obs}[Fe] \text{ where } k_{obs} = A[Y]/(B[X] + C[Y])$$

where [Fe] is shorthand for the reactant iron complex. It is thought that substitutions (15–10–29) occur by the dissociative mechanism in (15–10–31).

$$Fe(CN)_5X^{3-} \Leftrightarrow Fe(CN)_5^{3-} + X \; [k_1 \text{ (slow)}, k_{-1}\text{(fast)}]$$
$$Fe(CN)_5^{3-} + Y \Leftrightarrow Fe(CN)_5Y^{3-} \; [k_2 \text{ (fast)}, k_{-2}\text{(slow)}]$$

A steady-state assumption applied to the intermediate, $Fe(CN)_5^{3-}$, produces the rate law (15–10–32), which has the same form as the observed rate law.

$$\text{Rate} = (k_1k_2[Y] + k_{-1}k_{-2}[X])/(k_{-1}[X] + k_2[Y])\{[Fe]–[Fe]_{equil}\}$$

Because the reaction is reversible, the rate law is expressed in terms of the deviation of the concentration of $Fe(CN)_5X^{3-}$ from its equilibrium value. "Rate" then represents the rate of approach of the reaction to equilibrium.

By studying the manner in which k_{obs} varies with [X] and [Y], it is possible to assign values to the rate constants k_1, k_{-1}, k_2, and k_{-2}.

Substitution at Pt^{2+}. Most complexes of Pt^{2+} have four ligands arranged in a square planar arrangement around the Pt^{2+} ion. This leaves the Pt^{2+} center open to attack from above and below the plane of the complex, and suggests that an associative substitution mechanism might be possible.

FUNDAMENTAL SUBSTITUTION MECHANISMS

There are four mechanisms that have been proposed to account for the details of substitution reactions. These are presented below.

Concerted One-Step Substitution,

$$R–X + Y \rightarrow RY + X\ [k]$$

Rate law: Rate = k[RX][Y] (Second-order kinetics)

Associative Substitution,

$$R–X + Y \Leftrightarrow Y–R–X\ [k_1, k_{-1}]$$

$$Y–R–X \rightarrow R–Y + X\ [k_2]$$

$$\text{Rate} = k_1k_2[RX][Y]/(k_{-1} + k_2)\ \text{(Second-order kinetics)}$$

Dissociative Substitution,

$$R\text{-}X \Leftrightarrow R + X\ [k_1, k_{-1}]$$

$$R + Y \rightarrow RY\ [k_2]$$

$$\text{Rate} = k_1k_2[R–X][Y]/(k_{-1}[X] + k_2[Y])$$

Interchange,

$$R\text{-}X + Y \Leftrightarrow R\text{-}X...Y\ [K_1]$$

$$R–X...Y \rightarrow RY + X\ [k_2]$$

$$\text{Rate} = k_2K_1[R–X]_T[Y]/(1 + K_1[Y])$$

Supplement: Picosecond Reactions; Laser Promoted Processes.

Supplement: Kinetics, Mechanisms, and Occam's Razor.

APPLICATIONS

- Express the rate of the following reaction in terms of the change in concentration per unit time of each participating species:

$$N_2 + 3H_2 \rightarrow 2NH_3$$

- Complete the following statements for the reaction $2N_2O_5(g) \rightarrow 4NO_2(g) + O_2(g)$.
  - The rate of decomposition of N_2O_5 is _____ times the rate of

formation of O_2.

- The rate of formation of NO_2 is _____ times the rate of decomposition of N_2O_5.
- The rate of formation of NO_2 is _____ times the rate of formation of O_2.

- For the reaction in Application 1, what is the rate of appearance of ammonia if the rate of consumption of nitrogen is 0.01 moles L^{-1} s^{-1}? What is the reaction rate?
- In the reaction 3 $ClO^-(aq) \rightarrow 2\ Cl^-(aq) + ClO_3^-(aq)$, the rate of formation of Cl^- is 3.6 mol/L–min. What is the rate of reaction of ClO^-? What is the unique rate of reaction?
- Dinitrogen pentoxide, N_2O_5, decomposes by a first–order reaction. What is the initial rate of decomposition of N_2O_5 when 2.00 g of N_2O_5 is confined in a 1.00 L container and heated to 65°C? The rate constant for the reaction is $5.2 * 10^{-3}\ s^{-1}$ at 65°C.
- When the NO concentration is doubled, the rate of the reaction $2NO(g) + O_2(g) \rightarrow 2NO_2(g)$ increases by a factor of 4. When both the O_2 and NO concentrations are doubled, the rate increases by a factor of 8. What are the reactant orders and overall order of the reaction?
- Write the rate law for the consumption of persulfate ions in the reaction $S_2O_8^{2-}(aq) + 3I^-(aq) \rightarrow 2\ SO_4^{2-}(aq) + I_3^-(aq)$ with respect to each reactant and determine the value of k from the following data.

Experiment	Initial Concentration $S_2O_8^{2-}$ mol/L	Initial Concentration I^- mol/L	Initial Rate (mol $S_2O_8^{2-}$)$L^{-1}s^{-1}$
1	0.15	0.21	1.14
2	0.22	0.21	1.70
3	0.22	0.12	0.98

- The following kinetics data were obtained for the reaction, $2ICl(g) + H_2(g) \rightarrow I_2(g) + 2HCl(g)$.

Experiment	Initial Concentration ICl mmol/L	Initial Concentration H_2 mmol/L	Initial Rate mol $L^{-1}s^{-1}$
1	1.5	1.5	$3.7*10^{-7}$
2	3.0	1.5	$7.4*10^{-7}$
3	3.0	4.5	$22*10^{-7}$
4	4.7	2.7	?

- Write the rate law for the reaction.
- Determine the value of the rate constant from the data.
- Use the data to predict the reaction rate for experiment 4.

- The following data were obtained for the reaction A + B + C → products.

Experimen	$[A]_o$	$[B]_o$	$[C]_o$	Initial Rate mmol $L^{-1}s^{-1}$
1	1.25	1.25	1.25	8.7
2	2.5	1.25	1.25	17.4
3	1.25	3.02	1.25	50.8
4	1.25	3.02	3.75	457
5	3.01	1.00	1.15	?

- Write the rate laq for the reaction.
- What is the order of the reaction?
- Determine the value of the rate constant.
- Use the data to predict the reaction rate for experiment 5.

• Determine the rate constants for the following first–order reactions
  - A → B, given that the concentration of A decreases to one-half its initial value in 1000 s.
  - A → B, given that the concentration of A decreases from 0.33 mol*L^{-1} to 0.14 mol*L^{-1} in 47 s.
  - 2A → B + C, given that $[A]_o$ = 0.050 mol*L^{-1} and that after 120 s the concentration of B rises to 0.015 mol*L^{-1}.

• Dinitrogen pentoxide, N_2O_5, decomposes by first-order kinetics with a rate constant of 3.7 * 10^{-5} s^{-1} at 298 K.
  - What is the half-life in hours for the decomposition of N_2O_5 at 298 K?
  - If $[N_2O_5]_o$ = 2.33 * 10^{-2} mol*L^{-1}, what will the concentration of N_2O_5 be after 2.0 h?
  - How much time in minutes will elapse before the N_2O_5 concentration decreases from 2.33 * 10^{-2} mol*L^{-1} to 1.76 * 10^{-2} mol*L^{-1}?

• The half-life for the first-order decomposition of A is 200 s. How much time must elapse for the concentration of A to decrease to a) one-half, b) one–sixteenth, c) one-ninth of its initial concentration?

• Sulfuryl chloride, SO_2Cl_2, decomposes by first-order kinetics with k = 2.81 * 10^{-3} min^{-1} at a certain temperature.
  - What is the half-life for the reaction?
  - Determine the time needed for the concentration of a SO_2Cl_2 sample to decrease to 10% of its initial concentration.
  - If a 14.0–g sample of SO_2Cl_2 is sealed in a 2500–L reaction vessel and heated to the specified temperature, what mass will remain after 1.5 hr?

• For the first-order reaction A → 3B + C, when $[A]_o$ = 0.015 mol/L, the concentration of B increases to 0.020 mole/L in 3.0 min.
  - What is the rate constant for the reaction?

- How much more time would be needed for the concentration of B to increase to 0.040 mol/L?

- The following data were collected for the reaction $2N_2O_5(g) \rightarrow 4NO_2(g) + O_2(g)$ at 25 °C.

Time, s	**$[N_2O_5]$, 10^{-3} mol/L**
0	2.15
4000	1.8
8000	1.64
12000	1.43
16000	1.25

  - Plot the data appropriately to determine the order of reaction.
  - Determine the rate constant from the graph.

- The following data were collected for the reaction $2HI(g) \rightarrow H_2(g) + I_2(g)$ at 580 °C.

Time, s	**[HI], 10^{-3} mol/L**
0	1000
1000	112
2000	61
3000	41
4000	31

  - Plot the data appropriately to determine the reaction order.
  - Determine the rate constant from the graph.

- The half-life for the second-order reaction of a substance A is 50.5 s when $[A]_o$ = 0.84 mol/L. Calculate the time needed for the concentration of A to decrease to a) one-sixteenth; b) one-fourth; c) one-fifth of its original value.

- Determine the rate constant for the following second-order reactions
  - 2A ® B + 2C, given that the concentration of A decreases from 2.5 * 10^{-3} mol/L to 1.25 * 10^{-3} mole/L in 100 s.
  - 3A ® C + 2D, given that $[A]_o$ = 0.30 mol/L, and that the concentration of C increases to 0.010 mol/L in 200 s.

- Calculate the activation energy for the conversion of cyclopropane to propene from an Arrhenius plot of the following data:

T, K **k,** 750 1.8 * 00 2.7 * 10^{-3}

850 3.0 * 10^{-2}

900 0.26

What is the value of the rate constant at 600 °C?

- Write the overall reaction for the mechanism below, and identify intermediates:

$$ICl + H_2 \rightarrow HI + HCl$$
$$HI + ICl \rightarrow HCl + I_2$$

- Write the overall reaction for the mechanism below and identify intermediates:

$$Cl_2 \rightarrow 2Cl$$
$$Cl + CO \rightarrow COCl$$
$$COCl + Cl_2 \rightarrow COCl_2 + Cl$$

- Develop the rate law implied by the following mechanism:

$$NO + Br_2 \rightarrow NOBr_2 \text{ (slow)}$$
$$NOBr_2 + NO \rightarrow 2\ NOBr \text{ (fast)}$$

- Develop the rate law implied by the following mechanism:

$$Cl_2 \Leftrightarrow 2Cl \text{ (both fast, equilibrium)}$$
$$CHCl_3 + Cl \rightarrow CCl_3 + HCl \text{ (slow)}$$
$$CCl_3 + Cl \rightarrow CCl_4 \text{ (fast)}$$

- Develop the rate law implied by the following mechanism, proposed for the formation of phosgene from CO and Cl_2:

$$Cl_2 \Leftrightarrow 2\ Cl \text{ (both fast, equilibrium)}$$
$$Cl + CO \Leftrightarrow COCl \text{ (both fast, equilibrium)}$$
$$COCl + Cl_2 \rightarrow COCl_2 + Cl \text{ (slow)}$$

- Three mechanisms have been proposed for the reaction

$$NO_2(g) + CO(g) \rightarrow CO_2(g) + NO(g)$$

- *Mechanism 1:*
  - $NO_2 + CO \rightarrow NO + CO_2$
- *Mechanism 2:*
  - $NO_2 + NO_2$ ® $NO + NO_3$ (slow)
  - $NO_2 + NO_2$ Û $NO + NO_3$ (fast equil)
- *Mechanism 3:*
  - $NO_3 + CO$ ® $NO_2 + CO_2$ (fast)
  - $NO_3 + CO$ ® $NO_2 + CO_2$ (slow)

The observed rate law is Rate = $k[NO_2]^2$. Which mechanisms are feasible?

- For each mechanism in application 15–25, draw a reaction coordinate diagram.
- Critically analyse each of the following statements about elementary processes:
  - At equilibrium, the rate constants of the forward and reverse reactions are equal.
  - For a reaction with a very large equilibrium constant, the rate constant of the reverse reaction is much larger than the rate constant of the forward reaction.
  - The equilibrium constant for the reaction equals the ratio of the forward and reverse rates.
  - For an exothermic process, the rates of forward and reverse reactions are affected in the same way by a rise in temperature.

- For the reversible elementary process $2A \Leftrightarrow B + C$, the rate constant for the forward reaction is 265 $L^*mol^{-1}*min^{-1}$; the rate constant for the reverse reaction is 392 $L^*mol^{-1}*min^{-1}$; the activation energy for the forward reaction is 39.7 kJ/mol; and the activation energy for the reverse reaction is 25.4 kJ/mol.
  - Draw the reaction coordinate diagram.
  - What is the equilibrium constant for the reaction?
  - Is the reaction exo- or endothermic?
  - What effect will an increase in temperature have on the rate and equilibrium constants?
- For the reversible elementary process, $A + B \Leftrightarrow C + D$, the forward rate constant = 36.4 $L^*mol^{-1}*h^{-1}$; the reverse rate constant = 24.3 $L^*mol^{-1}*h^{-1}$; the forward activation energy is 33.8 kJ/mol; and the reverse activation energy is 45.4 kJ/mol.
  - Draw a reaction coordinate diagram.
  - What is K_{eq}?
  - Is the reaction exo- or endothermic?
  - What effect will an increase in temperature have on the rate and equilibrium constants?
- Suppose that a catalyst lowers E_a for a particular reaction from 100 to 50 kJ/mol. By what factor does the reaction rate increase at 400 K, all other factors being equal?
- A reaction rate increases by a factor of 10^3 in the presence of a catalyst at 2500°C. The activation energy of the uncatalysed path is 98 kJ/mol. What is the activation energy of the catalysed path, all other factors being equal?
- The reaction of nitric oxide and hydrogen,

$$2NO(g) + H_2(g) \rightarrow N_2O(g) + H_2O(g)$$

  is thought to proceed by the mechanism below.

$$1)\ 2NO \Leftrightarrow N_2O_2\ k_1, k_{-1}$$

$$2)\ N_2O_2 + H_2 \rightarrow N_2O + H_2O\ k_2$$

- The rate law for the reaction is Rate = $k[NO]^2$ at high pressure of H_2. At low H_2 pressure, the rate law is Rate = $k[H_2][NO]^2$. Apply the steady state approx-imation to the mechanism to obtain the full implied rate law, and show that it agrees with the experimental observations under the indicated conditions.
- For the second-order reaction $A \rightarrow B + C$, the concentration of A falls from 0.040 mol/L to 0.0050 mol/L in 12 h. What is the rate constant?
- All radioactive decay processes follow first-order kinetics. The half-life of the radioactive isotope tritium, ^3H or T, is 12.3 years. How much of a 1.0 mg ample of tritium would remain after 5.2 years?

- The rate law for the reaction $2NO(g) + 2H_2(g) \rightarrow N_2(g) + 2H_2O(g)$ is Rate = $k[NO]^2[H_2]$. The following mechanism has been proposed:

$$1)\ 2NO \rightarrow N_2O_2$$

$$2)\ N_2O_2 + H_2 \rightarrow N_2O + H_2O$$

$$3)\ N_2O + H_2 \rightarrow N_2 + H_2O$$

  - Which step is likely to be rate determining?
  - Sketch a reaction profile for the reaction, which is exothermic. Label the activation energy for each step and the overall reaction enthalpy.
- Use the following data to determine the activation energy for the reaction of ethyl bromide with hydroxide ions in aqueous solution.

Temperature, °C	**k, L/mole-s**
24	$1.3 * 10^{-3}$
28	$2.0 * 10^{-3}$
32	$3.0 * 10^{-3}$
26	$4.4 * 10^{-3}$
40	$6.4 * 10^{-3}$

- Explain in narrative form (*i.e.*, in words) why an enzyme catalyzed reaction approaches a maximum velocity when substrate concentration is steadily increased.
- The decarboxylation of a b-keto acid catalyzed by a decarboxylation enzyme can be measured by the rate of CO_2 formation. From the initial rates given in the table, determine the Michaelis-Menten constant for the enzyme and the maximum velocity by a graphical method.

[Keto Acid], M	Initial Velocity, **Micromoles CO_2/2 min**
2.500	0.588
1.000	0.500
0.714	0.417
0.526	0.370
0.250	0.256

- The hydration of CO_2,

$$CO_2 + H_2O \Leftrightarrow HCO_3^- + H^+$$

is catalyzed by the enzyme, carbonic anhydrase. The kinetics of the forward (hydration) and reverse (dehydration) reactions at pH 7.1, 0.5°C, and $2*10^{-3}$ M phosphate buffer were studied using bovine carbonic anhydrase. Some results follow:

Hydration	Dehydration		
1/v, $M^{-1}s$	**$[CO_2]$, mM**	**1/v, $M^{-1}s$**	**$[HCO_3^{-1}]$, mM**

$36*10^3$	1.25	$95*10^3$	2
$20*10^3$	2.5	$45*10^3$	5
$12*10^3$	5	$29*10^3$	10
$6*10^3$	20	$24*10^3$	15

Determine the Michaelis constant and the rate constant, k_2, for the decomposition of the enzyme–substrate complex to form product for:

- The hydration reaction
- The dehydration reaction
- From your results, determine the equilibrium constant for the reaction.

- At pH 7 the measured Michaelis constant and maximum velocity for the enzymatic conversion of fumarate to L-malate,

$$\text{Fumarate} + H_2O \rightarrow \text{L-malate}$$

are $4.0*10^{-6}$ M and $1.3*10^3[E]_o$ s^{-1}, respectively, where $[E]_o$ is the total molarity of the enzyme. The Michaelis constant and maximum velocity for the reverse reaction are $1.0*10^{-5}$ M and $800[E]_o$ s^{-1}, respectively. What is the equilibrium constant for the hydration reaction?

- The simple Michaelis-Menton mechanism for an enzmye catalyzed reaction is

$$E + S \Leftrightarrow ES, k_1, k_{-1}$$
$$ES \rightarrow E + P, k_2$$

The following data were obtained:

$$k_1, k_{-1} \text{ very fast}$$
$$k_2 = 100\ s^{-1}, K_M = 1.0*10^{-4} \text{ M at 280 K}$$
$$k_2 = 200\ s^{-1}, K_M = 1.5*10^{-4} \text{ M at 300 K}$$

  - For $[S] = 0.10$ M and $[E]_o = 1.0*10^{-5}$ M, calculate the rate of formation of product at 280 K.
  - Calculate the activation energy for k_2.
  - What is the equilibrium constant at 280 K for the formation of the enzyme substrate complex ES form E and S?
  - What is the sign and magnitude of DH° for the formation of ES from E and S?

- Explain in simple terms some of the reasons why enzymes are so *efficient* and so *specific*.
- A particular enzyme is present at a concentration of $1.2*10^{-8}$ M. The rate of substrate disappearance under a particular set of concentration and temperature conditions is 0.1 M/min. What is the turnover number of the enzyme?
- For the enzyme in the previous problem, an increase of pH by 1 unit slows the rate at which the enzyme converts substrate to

product by a substantial factor. What tentative conclusions can be drawn about the chemical environment at the active site of the enzyme?

- It is generally stated that a catalyst lowers the activation energy for a reaction by providing an alternative mechanism. Provided this is true, what general statement(s) can you make about the temperature dependence of the rate of an enzyme-catalyzed reaction versus that for the uncatalyzed reaction?
- Equation 11 of Kinetics Handout 2 shows the rate law for a reaction catalyzed by an enzyme in the presence of a *competitive inhibitor* of the enzyme. The rate law is obtained from a steady state treatment of the preceding mechanism. Carry out the steady state treatment and show that the rate law is correct.
- The basic mechanism for enzyme catalysis is usually written as if it involves only 2 steps: $E + S \Leftrightarrow ES$; $ES \rightarrow E + P$. The mechanism of action of chymotrypsin is discussed in Kinetics Handout 1. How many substeps are involved in the "second step" for chymotrypsin?
- Suppose that the equilibrium constant for binding of a molecule of substrate to a molecule of enzyme is $1.0*10^3$ M^{-1}. What fraction of the enzyme will be in the form of the enzyme–substrate complex when substrate concentration is 0.001 M. When substrate concentration is 0.01 M. By what factor will the substrate conversion rate increase when [S] is increased from 0.001 to 0.01 M?
- The equilibrium constant for binding of a molecule of substrate to a molecule of enzyme is 10^4 M^{-1}. What is the minimum substrate concentration necessary to guarantee that the enzyme is operating at more than 99% of its maximum conversion rate?
- You have obtained the following data for the destruction of thiamine by an enzyme, both with and without $0.20*10^{-4}$ M inhibitor.

[Thiamine], mM	Velocity (M/min), Uninhibited	Velocity (M/min), **Inhibited**
0.10	$0.746*10^{-6}$	$0.136*10^{-6}$
0.25	$0.992*10^{-6}$	$0.285*10^{-6}$
0.50	$1.12*10^{-6}$	$0.618*10^{-6}$
1.00	$1.26*10^{-6}$	$0.758*10^{-6}$
2.00	$1.36*10^{-6}$	$1.29*10^{-6}$

  - Make suitable plots and use the V_{max} and K_M values to determine the type of inhibition.
  - Based on these results, determine the binding constant for the inhibitor.

SPREADSHEET APPLICATIONS

- Carry out a finite difference analysis of the kinetics of the following sequence of elementary processes:

  $$A \rightarrow B \; k_1 = 0.20 \; s^{-1}$$
  $$B \rightarrow C \; k_2 = 0.15 \; s^{-1}$$

  Show how the concentrations of A, B, and C vary with time.
- The following concentration time data were obtained for the reaction

  $$A \rightarrow C$$

  Use finite difference methods to determine the order of the reaction in reactant A and the value of the rate constant.

Time, s	[A], M
0	0.15
40	0.127
80	0.107
120	0.0906
160	0.0766
200	0.0648
240	0.0547
280	0.0463
320	0.0391
360	0.0331
400	0.028
440	0.0236
48	0.020
520	0.0169
560	0.0143
600	0.0121
640	0.0102
680	0.00862
720	0.00729
760	0.00616
800	0.00521
840	0.0044
880	0.00372
920	0.00315
960	0.00266
1000	0.00225

- The following concentration time data were obtained for the reaction

  $$A + 2B \rightarrow C$$

  Use finite difference methods to determine the order of the reaction in each of the reactants A and B and the value of the rate constant.

Time, s	[A] for $[B]_o$ = 1.5 M	[A] for $[B]_o$ = 3.0 M
0	0.15	0.15

40	0.117	0.0906
80	0.0906	0.0547
120	0.0704	0.0331
160	0.0547	0.0200
200	0.0425	0.0121
240	0.0331	0.00729
280	0.0257	0.0044
320	0.0200	0.00266
360	0.0155	0.00161
400	0.0121	0.000971
440	0.00938	0.000587
480	0.00729	0.000354
520	0.00567	0.000214
560	0.00440	0.000129
600	0.00342	0.0000781
640	0.00266	0.0000472
680	0.002070	0000285
720	0.00161	0.0000172
760	0.00125	
800	0.000971	
840	0.000755	
880	0.000587	
920	0.000456	
960	0.000354	
1000	0.000275	

- At room temperature, A and B react to form two different products according to the following equations:

  $$A + B \rightarrow D \text{ with rate constant } k_1$$

  $$A + B \rightarrow F \text{ with rate constant } k_2$$

  It is possible to monitor the concentration of D at a wavelength of 540 nm, and the concentration of F at a wavelength of 280 nm. The follo194wing data were obtained in separate experiments, both run under pseudo-order conditions in A. What is the order of the reaction in A? What are the values of k_1 and k_2?

Time s	[D], M	[F], M
0	0.00E+00	0.00E+00
3	4.54E–06	1.34E–05
6	8.94E–06	2.55E–05
9	1.32E–05	3.64E–05
12	1.73E–05	4.63E–05
15	2.13E–05	5.53E–05
18	2.51E–05	6.34E–05
21	2.89E–05	7.07E–05
24	3.25E–05	7.73E–05
27	3.60E–05	8.33E–05
30	3.94E–05	8.88E–05
33	4.26E–05	9.37E–05

36	4.58E–05	9.81E–0
39	4.88E–05	1.02E–04
42	5.18E–05	1.06E–04
45	5.47E–05	1.09E–04
48	5.74E–05	1.12E–04
51	6.01E–05	1.15E–04
54	6.27E–05	1.17E–04
57	6.52E–05	1.19E–04
60	6.76E–05	1.21E–04
63	7.00E–05	1.23E–04
66	7.23E–05	1.25E–04

- The reaction A + B ⇔ D occurs as a reversible one-step process at room temperature. The kinetics of the net forward process were studied under pseudo-order conditions in A and the reaction was found to be first order in A. The observed rate constant, k_{obs}, was found to vary as follows with [B]. Determine the values of k_1 and k_{-1} for the reaction.

[A], M	**[B], M**	**k_{obs}, s^{-1}**
$1.2*10^{-4}$	0.124	$8.65*10^{-3}$
$1.2*10^{-4}$	0.273	$2.01*10^{-2}$
$1.2*10^{-4}$	0.314	$2.19*10^{-2}$
$1.8*10^{-4}$	0.116	$8.03*10^{-3}$
$1.9*10^{-4}$	0.437	$3.58*10^{-}$

6

The Water Cycle and Water Conservation

The water cycle, also known as the hydrologic cycle or H_2O cycle, describes the continuous movement of water on, above and below the surface of the Earth. Water can change states among liquid, vapour, and ice at various places in the water cycle. Although the balance of water on Earth remains fairly constant over time, individual water molecules can come and go, in and out of the atmosphere. The water moves from one reservoir to another, such as from river to ocean, or from the ocean to the atmosphere, by the physical processes of evaporation, condensation, precipitation, infiltration, run-off, and subsurface flow.

In so doing, the water goes through different phases: liquid, solid, and gas. The hydrologic cycle involves the exchange of heat energy, which leads to temperature changes. For instance, in the process of evaporation, water takes up energy from the surroundings and cools the environment. Conversely, in the process of condensation, water releases energy to its surroundings, warming the environment.

The water cycle figures significantly in the maintenance of life and ecosystems on Earth. Even as water in each reservoir plays an important role, the water cycle brings added significance to the presence of water on our planet. By transferring water from one reservoir to another, the water cycle purifies water, replenishes the land with freshwater, and transports minerals to different parts of the globe. It is also involved in reshaping the geological features of the Earth, through such processes as erosion and sedimentation. In addition, as the water cycle also involves heat exchange, it exerts an influence on climate as well.

DESCRIPTION

The sun, which drives the water cycle, heats water in oceans and seas. Water evaporates as water vapour into the air. Ice and snow can sublimate directly into water vapour. Evapotranspiration is water transpired from plants and evaporated from the soil. Rising air currents take the vapour up into the atmosphere where cooler temperatures cause it to condense into clouds. Air currents move water vapour around the globe, cloud particles collide, grow, and fall out of the sky as precipitation. Some precipitation falls as snow or hail, and can accumulate as ice caps and glaciers, which can store frozen water for thousands of years.

Snowpacks can thaw and melt, and the melted water flows over land as snowmelt. Most water falls back into the oceans or onto land as rain, where the water flows over the ground as surface run-off. A portion of run-off enters rivers in valleys in the landscape, with streamflow moving water towards the oceans. Run-off and groundwater are stored as freshwater in lakes.

Not all run-off flows into rivers, much of it soaks into the ground as infiltration. Some water infiltrates deep into the ground and replenishes aquifers, which store freshwater for long periods of time. Some infiltration stays close to the land surface and can seep back into surface-water bodies as groundwater discharge. Some groundwater finds openings in the land surface and comes out as freshwater springs. Over time, the water returns to the ocean, where our water cycle started.

PRECIPITATION

In meteorology, precipitation is any product of the condensation of atmospheric water vapour that falls under gravity. The main forms of precipitation include drizzle, rain, sleet, snow, graupel and hail. It occurs when a local portion of the atmosphere becomes saturated with water vapour and the water condenses Two processes, possibly acting together, can lead to air becoming saturated: cooling the air or adding water vapour to the air.

Generally, precipitation will fall to the surface; an exception is Virga which evaporates before reaching the surface. Precipitation forms as smaller droplets coalesce via collision with other rain drops or ice crystals within a cloud. Rain drops range in size from oblate, pancake-like shapes for larger drops, to small spheres for smaller drops. Unlike raindrops, snowflakes grow in a variety of different shapes and patterns, determined by the temperature and humidity characteristics of the air the snowflake moves through on its way to the ground. While snow and ice pellets require temperatures close to the ground to be near or below freezing, hail can occur during much warmer temperature regimes due to the process of its formation.

Moisture overriding associated with weather fronts is an overall major method of precipitation production. If enough moisture and upward motion is present, precipitation falls from convective clouds such as cumulonimbus and can organize into narrow rainbands. Where relatively warm water bodies are present, for example due to water evaporation from lakes, lake-effect snowfall becomes a concern downwind of the warm lakes within the cold cyclonic flow around the backside of extratropical cyclones. Lake-effect snowfall can be locally heavy. Thundersnow is possible within a cyclone's comma head and within lake effect precipitation bands. In mountainous areas, heavy precipitation is possible where upslope flow is maximized within windward sides of the terrain at elevation. On the leeward side of mountains, desert climates can exist due to the dry air caused by compressional heating. The movement of the monsoon trough, or intertropical convergence zone, brings rainy seasons to savannah climes.

Precipitation is a major component of the water cycle, and is responsible for depositing the fresh water on the planet. Approximately 505,000 cubic kilometres of water falls as precipitation each year; 398,000 cubic kilometres of it over the oceans. Given the Earth's surface area, that means the globally averaged annual precipitation is 990 millimetres. Climate classification systems such as the Köppen climate classification system use average annual rainfall to help differentiate between differing climate regimes. The urban heat island effect may lead to increased rainfall, both in amounts and intensity, downwind of cities. Global warming is also causing changes in the precipitation pattern globally. Precipitation may occur on other celestial bodies, *e.g.* when it gets cold, Mars has precipitation which most likely takes the form of ice needles, rather than rain or snow.

Hydrometeor

The term meteor describes an object from outer space which has entered the Earth's atmosphere and produces a light phenomenon. In contrast, any phenomenon which was at some point produced due to condensation or precipitation of moisture within the Earth's atmosphere is known as a hydrometeor. Particles composed of fallen precipitation which fell onto the Earth's surface can become hydrometeors if blown off the landscape by wind. Formations due to condensation such as clouds, haze, fog, and mist are composed of hydrometeors. All precipitation types are hydrometeors by definition, including virga, which is precipitation which evaporates before reaching the ground. Particles removed from the Earth's surface by wind such as blowing snow and blowing sea spray are also hydrometeors.

Types

Precipitation is a major component of the water cycle, and is responsible for depositing most of the fresh water on the planet. Approximately 505,000 km3 of water falls as precipitation each year, 398,000 km3 of it over the oceans. Given the Earth's surface area, that means the globally averaged annual precipitation is 990 millimetres. Mechanisms of producing precipitation include convective, stratiform, and orographic rainfall. Convective processes involve strong vertical motions that can cause the overturning of the atmosphere in that location within an hour and cause heavy precipitation, while stratiform processes involve weaker upward motions and less intense precipitation. Precipitation can be divided into three categories, based on whether it falls as liquid water, liquid water that freezes on contact with the surface, or ice. Mixtures of different types of precipitation, including types in different categories, can fall simultaneously. Liquid forms of precipitation include rain and drizzle. Rain or drizzle that freezes on contact within a subfreezing air mass is called "freezing rain" or "freezing drizzle". Frozen forms of precipitation include snow, ice needles, ice pellets, hail, and graupel.

How the Air becomes Saturated

Cooling Air to its Dew Point

The dew point is the temperature to which a parcel must be cooled in order to become saturated, and condenses to water. Water vapour normally begins to condense on condensation nuclei such as dust, ice, and salt in order to form clouds. An elevated portion of a frontal zone forces broad areas of lift, which form clouds decks such as altostratus or cirrostratus. Stratus is a stable cloud deck which tends to form when a cool, stable air mass is trapped underneath a warm air mass. It can also form due to the lifting of advection fog during breezy conditions.

There are four main mechanisms for cooling the air to its dew point: adiabatic cooling, conductive cooling, radiational cooling, and evaporative cooling. Adiabatic cooling occurs when air rises and expands. The air can rise due to convection, large-scale atmospheric motions, or a physical barrier such as a mountain. Conductive cooling occurs when the air comes into contact with a colder surface, usually by being blown from one surface to another, for example from a liquid water surface to colder land. Radiational cooling occurs due to the emission of infrared radiation, either by the air or by the surface underneath. Evaporative cooling occurs when moisture is added to the air through evaporation, which forces the air temperature to cool to its wet-bulb temperature, or until it reaches saturation.

Adding Moisture to the Air

The main ways water vapour is added to the air are: wind convergence into areas of upward motion, precipitation or virga falling from above, daytime heating evaporating water from the surface of oceans, water bodies or wet land, transpiration from plants, cool or dry air moving over warmer water, and lifting air over mountains.

Formation

Raindrops

Coalescence occurs when water droplets fuse to create larger water droplets, or when water droplets freeze onto an ice crystal, which is known as the Bergeron process. The fall rate of very small droplets is negligible, hence clouds do not fall out of the sky; precipitation will only occur when these coalesce into larger drops. When air turbulence occurs, water droplets collide, producing larger droplets. As these larger water droplets descend, coalescence continues, so that drops become heavy enough to overcome air resistance and fall as rain.

Raindrops have sizes ranging from 0.1 millimetres to 9 millimetres mean diameter, above which they tend to break up. Smaller drops are called cloud droplets, and their shape is spherical. As a raindrop increases in size, its shape becomes more oblate, with its largest cross-section facing the oncoming airflow.

Contrary to the cartoon pictures of raindrops, their shape does not resemble a teardrop. Intensity and duration of rainfall are usually inversely related, *i.e.*, high intensity storms are likely to be of short duration and low intensity storms can have a long duration. Rain drops associated with melting hail tend to be larger than other rain drops. The METAR code for rain is RA, while the coding for rain showers is SHRA.

Ice Pellets

Ice pellets or sleet are a form of precipitation consisting of small, translucent balls of ice. Ice pellets are usually smaller than hailstones. They often bounce when they hit the ground, and generally do not freeze into a solid mass unless mixed with freezing rain. The METAR code for ice pellets is PL.

Ice pellets form when a layer of above-freezing air exists with sub-freezing air both above and below it. This causes the partial or complete melting of any snowflakes falling through the warm layer. As they fall back into the sub-freezing layer closer to the surface, they re-freeze into ice pellets. However, if the sub-freezing layer beneath the warm layer is too small, the precipitation will not have time to re-freeze, and freezing rain will be the result at the surface. A temperature profile showing a warm layer above the ground is most likely to be found in advance of a warm front during the cold season, but can occasionally be found behind a passing cold front.

Hail

Like other precipitation, hail forms in storm clouds when supercooled water droplets freeze on contact with condensation nuclei, such as dust or dirt. The storm's updraft blows the hailstones to the upper part of the cloud. The updraft dissipates and the hailstones fall down, back into the updraft, and are lifted up again. Hail has a diameter of 5 millimetres or more. Within

METAR code, GR is used to indicate larger hail, of a diameter of at least 6.4 millimetres. GR is derived from the French word grêle. Smaller-sized hail, as well as snow pellets, use the coding of GS, which is short for the French word grésil. Stones just larger than golf ball-sized are one of the most frequently reported hail sizes. Hailstones can grow to 15 centimetres and weigh more than.5 kilograms.

In large hailstones, latent heat released by further freezing may melt the outer shell of the hailstone. The hailstone then may undergo 'wet growth', where the liquid outer shell collects other smaller hailstones. The hailstone gains an ice layer and grows increasingly larger with each ascent. Once a hailstone becomes too heavy to be supported by the storm's updraft, it falls from the cloud.

Snowflakes

Snow crystals form when tiny supercooled cloud droplets freeze. Once a droplet has frozen, it grows in the supersaturated environment. Because water droplets are more numerous than the ice crystals the crystals are able to grow to hundreds of micrometers or millimeters in size at the expense of the water droplets. This process is known as the Wegner-Bergeron-Findeison process. The corresponding depletion of water vapour causes the droplets to evaporate, meaning that the ice crystals grow at the droplets' expense.

These large crystals are an efficient source of precipitation, since they fall through the atmosphere due to their mass, and may collide and stick together in clusters, or aggregates. These aggregates are snowflakes, and are usually the type of ice particle that falls to the ground. Guinness World Records list the world's largest snowflakes as those of January 1887 at Fort Keogh, Montana; allegedly one measured 38 cm wide. The exact details of the sticking mechanism remain a subject of research.

Although the ice is clear, scattering of light by the crystal facets and hollows/imperfections mean that the crystals often appear white in Colour due to diffuse reflection of the whole spectrum of light by the small ice particles. The shape of the snowflake is determined broadly by the temperature and humidity at which it is formed. Rarely, at a temperature of around "2 °C, snowflakes can form in threefold symmetry—triangular snowflakes.

The most common snow particles are visibly irregular, although near-perfect snowflakes may be more common in pictures because they are more visually appealing. No two snowflakes are alike, which grow at different rates and in different patterns depending on the changing temperature and humidity within the atmosphere that the snowflake falls through on its way to the ground. The METAR code for snow is SN, while snow showers are coded SHSN.

Diamond Dust

Diamond dust, also known as ice needles or ice crystals, forms at temperatures approaching "40°F due to air with slightly higher moisture from aloft mixing with colder, surface based air. They are made of simple ice crystals that are hexagonal in shape. The METAR identifier for diamond dust within international hourly weather reports is IC.

Causes

Frontal Activity

Stratiform or dynamic precipitation occurs as a consequence of slow ascent of air in synoptic systems such as over surface cold fronts, and over and ahead of warm fronts. Similar ascent is seen around tropical cyclones outside of the eyewall, and in comma-head precipitation patterns around mid-latitude cyclones. A wide variety of weather can be found along an occluded front, with thunderstorms possible, but usually their passage is associated with a drying of the air mass. Occluded fronts usually form around mature low-pressure areas. Precipitation may occur on celestial bodies other than Earth. When it gets cold, Mars has precipitation that most likely takes the form of ice needles, rather than rain or snow.

Convection

Convective rain, or showery precipitation, occurs from convective clouds, *i.e.,* cumulonimbus or cumulus congestus. It falls as showers with rapidly changing intensity. Convective precipitation falls over a certain area for a relatively short time, as convective clouds have limited horizontal extent. Most precipitation in the tropics appears to be convective; however, it has been suggested that stratiform precipitation also occurs. Graupel and hail indicate convection. In mid-latitudes, convective precipitation is intermittent and often associated with baroclinic boundaries such as cold fronts, squall lines, and warm fronts.

Orographic Effects

Orographic precipitation occurs on the windward side of mountains and is caused by the rising air motion of a large-scale flow of moist air across the mountain ridge, resulting in adiabatic cooling and condensation. In mountainous parts of the world subjected to relatively consistent winds a more moist climate usually prevails on the windward side of a mountain than on the leeward or downwind side. Moisture is removed by orographic lift, leaving drier air on the descending and generally warming, leeward side where a rain shadow is observed. In Hawaii, on the island of Kauai, is notable for its extreme rainfall, as it has the second highest average annual rainfall on Earth, with 460 inches. Storm systems affect the state with heavy rains between October and March. Local climates vary considerably on each island due to their topography, divisible into windward and leeward regions based upon location relative to the higher mountains. Windward sides face the east to northeast trade winds and receive much more rainfall; leeward sides are drier and sunnier, with less rain and less cloud cover.

In South America, the Andes mountain range blocks Pacific moisture that arrives in that continent, resulting in a desertlike climate just downwind across western Argentina. The Sierra Nevada range creates the same effect in North America forming the Great Basin and Mojave Deserts.

Snow

Extratropical cyclones can bring cold and dangerous conditions with heavy rain and snow with winds exceeding 119 km/h. The band of precipitation that is associated with their warm front is often extensive, forced by weak upward vertical motion of air over the frontal boundary which condenses as it cools and produces precipitation within an elongated band, which is wide and stratiform, meaning falling out of nimbostratus clouds. When moist air tries to dislodge an arctic air mass, overrunning snow can result within the poleward side of the elongated precipitation band. In the Northern Hemisphere, poleward is towards the North Pole, or north. Within the Southern Hemisphere, poleward is towards the South Pole, or south.

Southwest of extratropical cyclones, curved cyclonic flow bringing cold air across the relatively warm water bodies can lead to narrow lake-effect snow bands. Those bands bring strong localized snowfall which can be understood as follows: Large water bodies such as lakes efficiently store heat that results in significant temperature differences between the water surface and the air above. Because of this temperature difference, warmth and moisture are transported upward, condensing into vertically oriented clouds which produce snow showers. The temperature decrease with height and cloud depth are directly affected by both the water temperature and the large-scale environment. The stronger the temperature decrease with height, the deeper the clouds get, and the greater the precipitation rate becomes.

In mountainous areas, heavy snowfall accumulates when air is forced to ascend the mountains and squeeze out precipitation along their windward slopes, which in cold conditions, falls in the form of snow. Because of the ruggedness of terrain, forecasting the location of heavy snowfall remains a significant challenge.

Within the Tropics

The wet, or rainy, season is the time of year, covering one or more months, when most of the average annual rainfall in a region falls. The term green season is also sometimes used as a euphemism by tourist authorities. Areas with wet seasons are dispersed across portions of the tropics and subtropics. Savanna climates and areas with monsoon regimes have wet summers and dry winters. Tropical rainforests technically do not have dry or wet seasons, since their rainfall is equally distributed through the year. Some areas with pronounced rainy seasons will see a break in rainfall mid-season when the intertropical convergence zone or monsoon trough move poleward of their location during the middle of the warm season.

When the wet season occurs during the warm season, or summer, rain falls mainly during the late afternoon and early evening hours. The wet season is a time when air quality improves, freshwater quality improves, and vegetation grows significantly. Soil nutrients diminish and erosion increases. Animals have adaptation and survival strategies for the wetter regime. Tropical cyclones, a source of very heavy rainfall, consist of large air masses several hundred miles across with low pressure at the centre and with winds blowing inward towards the centre in either a clockwise direction or counterclockwise. Although cyclones can take an enormous toll in lives and personal property, they may be important factors in the precipitation regimes of places they impact, as they may bring much-needed precipitation to otherwise dry regions. Areas in their path can receive a year's worth of rainfall from a tropical cyclone passage.

Large-scale Geographical Distribution

On the large scale, the highest precipitation amounts outside topography fall in the tropics, closely tied to the Intertropical Convergence Zone, itself the ascending branch of the Hadley cell. Mountainous locales near the equator in Colombia are amongst the wettest places on Earth. North and south of this are regions of descending air that form subtropical ridges where precipitation is low; the land surface underneath is usually arid, which forms most of the Earth's deserts. An exception to this rule is in Hawaii, where upslope flow due to the trade winds lead to one of the wettest locations on Earth. Otherwise, the flow of the Westerlies into the Rocky Mountains lead to the wettest, and at elevation snowiest, locations within North America. In Asia during the wet season, the flow of moist air into the Himalayas leads to some of the greatest rainfall amounts measured on Earth in northeast India.

Measurement

The standard way of measuring rainfall or snowfall is the standard rain gauge, which can be found in 100 mm plastic and 200 mm metal varieties. The inner cylinder is filled by 25 mm of rain, with overflow flowing into the outer cylinder. Plastic gauges have markings on the inner cylinder down to 0.25 mm resolution, while metal gauges require use of a stick designed with the appropriate 0.25 mm markings. After the inner cylinder is filled, the amount inside it is discarded, then filled with the remaining rainfall in the outer cylinder until all the fluid in the outer cylinder is gone, adding to the overall total until the outer cylinder is empty. These gauges are used in the winter by removing the funnel and inner cylinder and allowing snow and freezing rain to collect inside the outer cylinder. Some add anti-freeze to their gauge so they do not have to melt the snow or ice that falls into the gauge. Once the snowfall/ice is finished accumulating, or as 300 mm is approached, one can either bring it inside to melt, or use lukewarm water to fill the inner cylinder with in order to melt the frozen precipitation in the outer cylinder, keeping track of the warm fluid added, which is subsequently subtracted from the overall total once all the ice/snow is melted.

Other types of gauges include the popular wedge gauge, the tipping bucket rain gauge, and the weighing rain gauge. The wedge and tipping bucket gauges will have problems with snow. Attempts to compensate for snow/ice by warming the tipping bucket meet with limited success, since snow may sublimate if the gauge is kept much above freezing. Weighing gauges with antifreeze should do fine with snow, but again, the funnel needs to be removed before the event begins. For those looking to measure rainfall the most inexpensively, a can that is cylindrical with straight sides will act as a rain gauge if left out in the open, but its accuracy will depend on what ruler is used to measure the rain with. Any of the above rain gauges can be made at home, with enough know-how.

When a precipitation measurement is made, various networks exist across the United States and elsewhere where rainfall measurements can be submitted through the Internet, such as CoCoRAHS or GLOBE. If a network is not available in the area where one lives, the nearest local weather office will likely be interested in the measurement.

Return Period

The likelihood or probability of an event with a specified intensity and duration, is called the return period or frequency. The intensity of a storm can be predicted for any return period and storm duration, from charts based on historic data for the location. The term 1 in 10 year storm describes a rainfall event which is rare and is only likely to occur once every 10 years, so it has a 10 per cent likelihood any given year. The rainfall will be greater and the flooding will be worse than the worst storm expected in any single year. The term 1 in 100 year storm describes a rainfall event which is extremely rare and which will occur with a likelihood of only once in a century, so has a 1 per cent likelihood in any given year. The rainfall will be extreme and flooding to be worse than a 1 in 10 year event. As with all probability events, it is possible to have multiple "1 in 100 Year Storms" in a single year.

Role in Köppen Climate Classification

The Köppen classification depends on average monthly values of temperature and precipitation. The most commonly used form of the Köppen classification has five primary types labeled A through E. Specifically, the primary types are A, tropical; B, dry; C, mild mid-latitude; D, cold mid-latitude; and E, polar. The five primary classifications can be further divided into secondary classifications such as rain forest, monsoon, tropical savanna, humid subtropical, humid continental, oceanic climate, Mediterranean climate, steppe, subarctic climate, tundra, polar ice cap, and desert.

Rain forests are characterized by high rainfall, with definitions setting minimum normal annual rainfall between 1,750 millimetres and 2,000 millimetres. A tropical savanna is a grassland biome located in semi-arid to semi-humid climate regions of subtropical and tropical latitudes, with rainfall between 750 millimetres and 1,270 millimetres a year. They are widespread on Africa, and are also found in India, the northern parts of South America, Malaysia, and Australia. The humid subtropical climate zone where winter rainfall is associated with large storms that the westerlies steer from west to east. Most summer rainfall occurs during thunderstorms and from occasional tropical cyclones. Humid subtropical climates lie on the east side continents, roughly between latitudes 20° and 40° degrees away from the equator.

An oceanic climate is typically found along the west coasts at the middle latitudes of all the world's continents, bordering cool oceans, as well as southeastern Australia, and is accompanied by plentiful precipitation year round. The Mediterranean climate regime resembles the climate of the lands in the Mediterranean Basin, parts of western North America, parts of Western and South Australia, in southwestern South Africa and in parts of central Chile. The climate is characterized by hot, dry summers and cool, wet winters. A steppe is a dry grassland. Subarctic climates are cold with continuous permafrost and little precipitation.

Effect on Agriculture

Precipitation, especially rain, has a dramatic effect on agriculture. All plants need at least some water to survive, therefore rain is important to agriculture. While a regular rain pattern is usually vital to healthy plants, too much or too little rainfall can be harmful, even devastating to crops. Drought can kill crops and increase erosion, while overly wet weather can cause harmful fungus growth. Plants need varying amounts of rainfall to survive. For example, certain cacti require small amounts of water, while tropical plants may need up to hundreds of inches of rain per year to survive.

In areas with wet and dry seasons, soil nutrients diminish and erosion increases during the wet season. Animals have adaptation and survival strategies for the wetter regime. The previous dry season leads to food shortages into the wet season, as the crops have yet to mature. Developing countries have noted that their populations show seasonal weight fluctuations due to food shortages seen before the first harvest, which occurs late in the wet season.

Changes Due to Global Warming

Increasing temperatures tend to increase evaporation which leads to more precipitation. Precipitation has generally increased over land north of 30°N from 1900 through 2005 but has declined over the tropics since the 1970s. Globally there has been no statistically significant overall trend in precipitation over the past century, although trends have varied widely by region and over time. Eastern portions of North and South America, northern Europe, and northern and central Asia have become wetter.

The Sahel, the Mediterranean, southern Africa and parts of southern Asia have become drier. There has been an increase in the number of heavy precipitation events over many areas during the past century, as well as an increase since the 1970s in the prevalence of droughts—especially in the tropics and subtropics. Changes in precipitation and evaporation over the oceans are suggested by the decreased salinity of mid-and high-latitude waters along with increased salinity in lower latitudes. Over the contiguous United States, total annual precipitation increased at an average rate of 6.1 per cent per century since 1900, with the greatest increases within the East North Central climate region and the South. Hawaii was the only region to show a decrease.

Changes Due to Urban Heat Island

The urban heat island warms cities 0.6 °C to 5.6 °C above surrounding suburbs and rural areas. This extra heat leads to greater upward motion, which can induce additional shower and thunderstorm activity. Rainfall rates downwind of cities are increased between 48% and 116%. Partly as a result of this warming, monthly rainfall is about 28% greater between 20 miles to 40 miles downwind of cities, compared with upwind. Some cities induce a total precipitation increase of 51%.

Forecasting

The Quantitative Precipitation Forecast is the expected amount of liquid precipitation accumulated over a specified time period over a specified area.

A QPF will be specified when a measurable precipitation type reaching a minimum threshold is forecast for any hour during a QPF valid period. Precipitation forecasts tend to be bound by synoptic hours such as 0000, 0600, 1200 and 1800 GMT. Terrain is considered in QPFs by use of topography or based upon climatological precipitation patterns from observations with fine detail. Starting in the mid to late 1990s, QPFs were used within hydrologic forecast models to simulate impact to rivers throughout the United States. Forecast models show significant sensitivity to humidity levels within the planetary boundary layer, or in the lowest levels of the atmosphere, which decreases with height. QPF can be generated on a quantitative, forecasting amounts, or a qualitative, forecasting the probability of a specific amount, basis. Radar imagery forecasting techniques show higher skill than model forecasts within six to seven hours of the time of the radar image. The forecasts can be verified through use of rain gage measurements, weather radar estimates, or a combination of both. Various skill scores can be determined to measure the value of the rainfall forecast.

TRANSPIRATION

Transpiration is a process similar to evaporation. It is a part of the water cycle, and it is the loss of water vapour from parts of plants, especially in leaves but also in stems, flowers and roots. Leaf surfaces are dotted with openings which are collectively called stomata, and in most plants they are more numerous on the undersides of the foliage. The stoma are bordered by guard cells that open and close the pore. Leaf transpiration occurs through stomata, and can be thought of as a necessary "cost" associated with the opening of the stomata to allow the diffusion of carbon dioxide gas from the air for photosynthesis. Transpiration also cools plants and enables mass flow of mineral nutrients and water from roots to shoots.

Mass flow of liquid water from the roots to the leaves is caused by the decrease in hydrostatic pressure in the upper parts of the plants due to the diffusion of water out of stomata into the atmosphere. Water is absorbed at the roots by osmosis, and any dissolved mineral nutrients travel with it through the xylem.

The rate of transpiration is directly related to the evaporation of water molecules from plant surface, especially from the surface openings, or stoma, on leaves. Stomatal transpiration accounts for most of the water loss by a plant, but some direct evaporation also takes place through the cuticle of the leaves and young stems. The amount of water given off depends somewhat upon how much water the roots of the plant have absorbed.

It also depends upon such environmental conditions as light intensity, humidity, winds and temperature. A plant should not be transplanted in full sunshine because it may lose too much water and wilt before the damaged roots can supply enough water. Transpiration occurs as the sun warms the water inside the blade. The warming changes much of the water into water vapour. This gas can then escape through the stomata. Transpiration helps cool the inside of the leaf because the escaping vapour has absorbed heat, the degree of stomatal opening, and the evaporative demand of the atmosphere surrounding the leaf. The amount of water lost by a plant depends on its size, along with surrounding light intensity, temperature, humidity, and wind speed. Soil water supply and soil temperature can influence stomatal opening, and thus transpiration rate. A fully grown tree may lose several hundred gallons of water through its leaves on a hot, dry day. About 90% of the water that enters a plant's roots is used for this process. The transpiration ratio is the ratio of the mass of water transpired to the mass of dry matter produced; the transpiration ratio of crops tends to fall between 200 and 1000. Transpiration rate of plants can be measured by a number of techniques, including potometers, lysimeters, porometers, photosynthesis systems and heat balance sap flow gauges.

Desert plants and conifers have specially adapted structures, such as thick cuticles, reduced leaf areas, sunken stomata and hairs to reduce transpiration and conserve water. Many cacti conduct photosynthesis in succulent stems, rather than leaves, so the surface area of the shoot is very low. Many desert plants have a special type of photosynthesis, termed crassulacean acid metabolism or CAM photosynthesis in which the stomata are closed during the day and open at night when transpiration will be lower.

CONDENSATION

Condensation is the change of the physical state of matter from gaseous phase into liquid phase, and is the reverse of evaporation. When the transition happens from the gaseous phase into the solid phase directly, the change is called deposition.

Upon the slowing-down of the molecules of the material, the overall attraction forces between these prevail and bring them together at distances comparable to their sizes. Since the condensing molecules suffer from reduced degrees of freedom and ranges of motion, their prior kinetic energy must be transferred to an absorbing colder entity—either a center of condensation within the gas volume or some contact surface. Condensation is initiated by the formation of atomic/molecular clusters of that species within its gaseous volume—like rain drop or snow-flake formation within clouds—or at the contact between such gaseous phase and a liquid or solid surface.

A few distinct reversibility scenarios emerge here with respect to the nature of the surface:

- Absorption into the surface of a liquid is reversible as evaporation.

- Adsorption onto solid surface at pressures and temperatures higher than the specie's triple point—also reversible as evaporation.
- Adsorption onto solid surface at pressures and temperatures lower than the specie's triple point—is reversible as sublimation.

Condensation commonly occurs when a vapour is cooled and/or compressed to its saturation limit when the molecular density in the gas phase reaches its maximal threshold. Vapour cooling and compressing equipment that collects condensed liquids is called "condenser".

Psychrometry measures the rates of condensation from and evaporation into the air moisture at various atmospheric pressures and temperatures. Water is the product of its vapour condensation—condensation is the process of such phase conversion.

Applications of Condensation

Condensation is a crucial component of distillation, an important laboratory and industrial chemistry application. Because condensation is a naturally occurring phenomenon, it can often be used to generate water in large quantities for human use. Many structures are made solely for the purpose of collecting water from condensation, such as air wells and fog fences. Such systems can often be used to retain soil moisture in areas where active desertification is occurring—so much so that some organizations educate people living in affected areas about water condensers to help them deal effectively with the situation.

It is also a crucial process in forming particle tracks in a Cloud Chamber. In this case, ions produced by an incident particle act as nucleation centres for the vapour to condense on.

Biological Adaptation

Numerous living organisms use water made accessible by condensation. A few examples are the Australian Thorny Devil, Darkling beetles on the Namibian coast and Coast Redwoods on the west coast of the United States.

ADVECTION

Advection, in chemistry, engineering and earth sciences, is a transport mechanism of a substance, or a conserved property, by a fluid, due to the fluid's bulk motion in a particular direction. An example of advection is the transport of pollutants or silt in a river. The motion of the water carries these impurities downstream. Another commonly advected property is energy or enthalpy, and here the fluid may be water, air, or any other thermal energy-containing fluid material. Any substance, or conserved property can be advected, in a similar way, in any fluid. The fluid motion in advection is described mathematically as a vector field, and the material transported is typically described as a scalar concentration of substance, which is contained in the fluid. Advection requires currents in the fluid, and so cannot happen in solids.

It does not include transport of substances by simple diffusion. Advection is sometimes confused with the more encompassing process convection, which encompasses both advective transport and diffusive transport in fluids. Convective transport is the sum of advective transport and diffusive transport.

Advection is important for the formation of orographic cloud and the precipitation of water from clouds, as part of the hydrological cycle. In meteorology and physical oceanography, advection often refers to the transport of some property of the atmosphere or ocean, such as heat, humidity or salinity. Meteorological or oceanographic advection follows isobaric surfaces and is therefore predominantly horizontal.

Distinction between Advection and Convection

Occasionally, the term advection is used as synonymous with convection. However, many engineers prefer to use the term convection to describe transport by combined molecular and eddy diffusion, and reserve the usage of the term advection to describe transport with a general flow of the fluid.

An example of convection is flow over a hot plate or below a chilled water surface in a lake. In the ocean and atmospheric sciences, advection is understood as horizontal movement resulting in transport "from place to place", while convection is vertical "mixing".

Meteorology

In meteorology and physical oceanography, advection often refers to the transport of some property of the atmosphere or ocean, such as heat, humidity or salinity. Advection is important for the formation of orographic clouds and the precipitation of water from clouds, as part of the hydrological cycle.

Other Quantities

The advection equation also applies if the quantity being advected is represented by a probability density function at each point, although accounting for diffusion is more difficult.

Mathematics of Advection

The advection equation is the partial differential equation that governs the motion of a conserved scalar as it is advected by a known velocity field. It is derived using the scalar's conservation law, together with Gauss's theorem, and taking the infinitesimal limit.

Perhaps the best image to have in mind is the transport of salt dumped in a river. If the river is originally fresh water and is flowing quickly, the predominant form of transport of the salt in the water will be advective, as the water flow itself would transport the salt. If the river was not flowing the salt would simply disperse outwards from its source in a diffusive manner, which is not advection.

In Cartesian coordinates the advection operator is,

$$u.\nabla = u\frac{\partial}{\partial x} + v\frac{\partial}{\partial z}$$

where the velocity vector u has components u, v and w in the x, y and z directions respectively.

The advection equation for a scalar ψ, such as temperature, is expressed mathematically as:

$$\frac{\partial \psi}{\partial} + \nabla.(\psi u) = 0$$

where ∇ is the divergence operator and u is the velocity vector field. Frequently, it is assumed that the flow is incompressible, that is, the velocity field satisfies $\nabla \cdot u = 0$ If this is so, the above equation reduces to,

$$\frac{\partial \psi}{\partial t} + u \cdot \nabla \psi = 0.$$

For a vector a, such as magnetic field or velocity, in a solenoidal field it is defined as:

$$\frac{\partial a}{\partial t} + (u \cdot \nabla)a = 0.$$

In particular, if the flow is steady, $u \cdot \nabla \psi = 0$ which shows that ψ is constant along a streamline.

The advection equation is not simple to solve numerically: the system is a hyperbolic partial differential equation, and interest typically centers on discontinuous "shock" solutions. Even in one space dimension and constant velocity, the system remains difficult to simulate.

The equation becomes,

$$\frac{\partial \psi}{\partial t} + u\frac{\partial \psi}{\partial x} = 0$$

where ψ = ψ(x,t) is the scalar being advected and u the x component of the vector $u = (u(x),0,0)$.

Numerical simulation can be aided by considering the skew symmetric form for the advection operator.

$$\frac{1}{2}u \cdot \nabla u + \frac{1}{2}\nabla(uu)$$

where ∇(uu) is a vector with components,

$$\left[\nabla\left(uu_y\right),\nabla\left(uu_y\right),\nabla\left(uu_z\right)\right]$$

and the notation $u=\left[u_z,u_y,u_z\right]$ has been used.

Since skew symmetry implies only imaginary eigenvalues, this form reduces the "blow up" and "spectral blocking" often experienced in numerical solutions with sharp discontinuities. Using vector calculus identities, these operators can also be expressed in other ways, available in more software packages for more coordinate systems,

$$u\cdot\nabla u=\nabla\left(\frac{\|u\|^2}{2}\right)+(\nabla\times u)\times u$$

$$\frac{1}{2}u\cdot\nabla u+\frac{1}{2}\nabla(uu)=\nabla\left(\frac{\|u\|^2}{2}\right)+(\nabla\times u)\times u+\frac{1}{2}u(\nabla\cdot u)$$

This form also makes visible that the skew symmetric operator introduces error when the velocity field diverges.

SUBLIMATION

Sublimation refers to the process of transition of a substance from the solid phase to the gas phase without passing through an intermediate liquid phase. Sublimation is an endothermic phase transition that occurs at temperatures and pressures below a substance's triple point in its phase diagram. At normal pressures, most chemical compounds and elements possess three different states at different temperatures. In these cases the transition from the solid to the gaseous state requires an intermediate liquid state. Note, however, that the pressure referred to here is the partial pressure of the substance, not the total pressure of the entire system. So, all solids which possess an appreciable vapour pressure at a certain temperature usually can sublime in air.

For some substances, such as carbon and arsenic, sublimation is much easier than evaporation from the melt, because the pressure of their triple point is very high, and it is difficult to obtain them as liquids. Sublimation requires additional energy and is an endothermic change. The enthalpy of sublimation can be calculated as the enthalpy of fusion plus the enthalpy of vapourization. The reverse process of sublimation is deposition. The formation of frost is an example of meteorological deposition.

Examples

- *Carbon dioxide*: Solid carbon dioxide sublimes readily at atmospheric pressure at–78.5°C, while liquid CO_2 can be obtained at pressures and temperatures above the triple point.

- *Water*: Snow and ice sublime, although more slowly, below the melting point temperature. This allows wet cloth to be hung outdoors in freezing weather and retrieved later in a dry state. In freeze–drying the material to be dehydrated is frozen and its water is allowed to sublime under reduced pressure or vacuum. The loss of snow from a snowfield during a cold spell is often caused by sunshine acting directly on the upper layers of the snow. Ablation is a process which includes sublimation and erosive wear of glacier ice.
- *Other compounds*: Iodine produces fumes on gentle heating. It is possible to obtain liquid iodine at atmospheric pressure by controlling the temperature at just above the melting point of iodine. Naphthalene, a common ingredient in mothballs, also sublimes easily. Arsenic can also sublime at high temperatures. Various substances appear to sublime because of undergoing chemical reactions or decomposition; for example, ammonium chloride when heated decomposes into hydrogen chloride and ammonia.

Sublimation Purification

Sublimation is a technique used by chemists to purify compounds. Typically a solid is placed in a sublimation apparatus and heated under vacuum. Under this reduced pressure the solid volatilizes and condenses as a purified compound on a cooled surface, leaving a non-volatile residue of impurities behind. Once heating ceases and the vacuum is removed, the purified compound may be collected from the cooling surface.

Historical Usage

In alchemy, sublimation typically referred to the process by which a substance is heated to a vapour, then immediately collects as sediment on the upper portion and neck of the heating medium. It is one of the 12 core alchemical processes.

EVAPORATION

Evaporation is a type of vapourization of a liquid that occurs only on the surface of a liquid. The other type of vapourization is boiling, which, instead, occurs on the entire mass of the liquid. Evaporation is also part of the water cycle. On average, the molecules in a glass of water do not have enough heat energy to escape from the liquid. With sufficient heat, the liquid would turn into vapour quickly. When the molecules collide, they transfer energy to each other in varying degrees, based on how they collide. Sometimes the transfer is so one-sided for a molecule near the surface that it ends up with enough energy to escape. Liquids that do not evaporate visibly at a given temperature in a given gas have molecules that do not tend to transfer energy to each other in a pattern sufficient to frequently give a molecule the heat energy necessary to turn into vapour. However, these liquids are evaporating. It is just that the process is much slower and thus significantly less visible.

Evaporation is an essential part of the water cycle. Solar energy drives evaporation of water from oceans, lakes, moisture in the soil, and other sources of water. In hydrology, evaporation and transpiration are collectively termed evapotranspiration. Evaporation is caused when water is exposed to air and the liquid molecules turn into water vapour, which rises up and forms clouds. The temperature that evaporation occurs ranges from 8.33°F-107.33°F

Theory

For molecules of a liquid to evaporate, they must be located near the surface, be moving in the proper direction, and have sufficient kinetic energy to overcome liquid-phase intermolecular forces. Only a small proportion of the molecules meet these criteria, so the rate of evaporation is limited. Since the kinetic energy of a molecule is proportional to its temperature, evaporation proceeds more quickly at higher temperatures. As the faster-moving molecules escape, the remaining molecules have lower average kinetic energy, and the temperature of the liquid, thus, decreases. This phenomenon is also called evaporative cooling. This is why evaporating sweat cools the human body. Evaporation also tends to proceed more quickly with higher flow rates between the gaseous and liquid phase and in liquids with higher vapour pressure. For example, laundry on a clothes line will dry more rapidly on a windy day than on a still day. Three key parts to evaporation are heat, humidity, and air movement.

On a molecular level, there is no strict boundary between the liquid state and the vapour state. Instead, there is a Knudsen layer, where the phase is undetermined. Because this layer is only a few molecules thick, at a macroscopic scale a clear phase transition interface can be seen.

Evaporative Equilibrium

If evaporation takes place in a closed vessel, the escaping molecules accumulate as a vapour above the liquid. Many of the molecules return to the liquid, with returning molecules becoming more frequent as the density and pressure of the vapour increases. When the process of escape and return reaches an equilibrium, the vapour is said to be "saturated," and no further change in either vapour pressure and density or liquid temperature will occur. For a system consisting of vapour and liquid of a pure substance, this equilibrium state is directly related to the vapour pressure of the substance, as given by the Clausius-Clapeyron relation:

$$\ln\left(\frac{P_2}{P_1}\right) = \frac{\Delta H_{vap}}{R}\left(\frac{1}{T_2} - \frac{1}{T_1}\right)$$

where P_1, P_2 are the vapour pressures at temperatures T_1, T_2 respectively, ΔH_{vap} is the enthalpy of vapourization, and R is the universal gas constant. The rate of evaporation in an open system is related to the vapour pressure found in a closed system.

If a liquid is heated, when the vapour pressure reaches the ambient pressure the liquid will boil. The ability for a molecule of a liquid to evaporate is based largely on the amount of kinetic energy an individual particle may possess. Even at lower temperatures, individual molecules of a liquid can evaporate if they have more than the minimum amount of kinetic energy required for vapourization.

Factors Influencing the Rate of Evaporation

- *Concentration of the substance evaporating in the air*: If the air already has a high concentration of the substance evaporating, then the given substance will evaporate more slowly.
- *Concentration of other substances in the air*: If the air is already saturated with other substances, it can have a lower capacity for the substance evaporating.
- *Concentration of other substances in the liquid*: If the liquid contains other substances, it will have a lower capacity for evaporation.
- *Flow rate of air*: This is in part related to the concentration points above. If fresh air is moving over the substance all the time, then the concentration of the substance in the air is less likely to go up with time, thus encouraging faster evaporation. This is the result of the boundary layer at the evaporation surface decreasing with flow velocity, decreasing the diffusion distance in the stagnant layer.
- *Inter-molecular forces*: The stronger the forces keeping the molecules together in the liquid state, the more energy one must get to escape. This is characterized by the enthalpy of vapourization.
- *Pressure*: Evaporation happens faster if there is less exertion on the surface keeping the molecules from launching themselves.
- *Surface area*: A substance that has a larger surface area will evaporate faster, as there are more surface molecules that are able to escape.

- *Temperature of the substance*: If the substance is hotter, then its molecules have a higher average kinetic energy, and evaporation will be faster.
- *Density*: The higher the density the slower a liquid evaporates.

In the US, the National Weather Service measures the actual rate of evaporation from a standardized "pan" open water surface outdoors, at various locations nationwide. Others do likewise around the world. The US data is collected and compiled into an annual evaporation map. The measurements range from under 30 to over 120 inches per year.

Applications

- When clothes are hung on a laundry line, even though the ambient temperature is below the boiling point of water, water evaporates. This is accelerated by factors such as low humidity, heat and wind. In a clothes dryer, hot air is blown through the clothes, allowing water to evaporate very rapidly.
- The botijo, a traditional Spanish porous clay container designed to cool the contained water by evaporation.

SUBSURFACE FLOW

Subsurface flow, in hydrology, is the flow of water beneath earth's surface as part of the water cycle.

In the water cycle, when precipitation falls on the earth's land, some of the water flows on the surface forming streams and rivers. The remaining water, through infiltration, penetrates the soil traveling underground, hydrating the vadose zone soil, recharging aquifers, with the excess flowing in subsurface run-off. In hydrogeology it is measured by the Groundwater flow equation.

Run-off

Water flows from areas where the water table is higher to areas where it is lower. This flow can be either surface run-off in rivers and streams, or subsurface run-off infiltrating rocks and soil. The amount of run-off reaching surface and groundwater can vary significantly, depending on rainfall, soil moisture, permeability, groundwater storage, evaporation, upstream use, and whether or not the ground is frozen. The movement of subsurface water is determined largely by the water gradient, type of substrate, and any barriers to flow.

Surface Return

Subsurface water may return to the surface in groundwater flow, such as from a spring, seep, or a water well, or subsurface return to streams, rivers, and oceans. Water returns to the land surface at a lower elevation than where infiltration occurred, under the force of gravity or gravity induced pressures. Groundwater tends to move slowly, and is replenished slowly, so it can remain in aquifers for thousands of years. Mainly, water flows through the ground which leads to the ocean where the cycle begins again.

INFILTRATION

Infiltration is the process by which water on the ground surface enters the soil. Infiltration rate in soil science is a measure of the rate at which soil is able to absorb rainfall or irrigation. It is measured in inches per hour or millimeters per hour.

The rate decreases as the soil becomes saturated. If the precipitation rate exceeds the infiltration rate, run-off will usually occur unless there is some physical barrier. It is related to the saturated hydraulic conductivity of the near-surface soil. The rate of infiltration can be measured using an infiltrometer.

Infiltration is governed by two forces: gravity and capillary action. While smaller pores offer greater resistance to gravity, very small pores pull water through capillary action in addition to and even against the force of gravity.

The rate of infiltration is affected by soil characteristics including ease of entry, storage capacity, and transmission rate through the soil.

The soil texture and structure, vegetation types and cover, water content of the soil, soil temperature, and rainfall intensity all play a role in controlling infiltration rate and capacity. For example, coarse-grained sandy soils have large spaces between each grain and allow water to infiltrate quickly. Vegetation creates more porous soils by both protecting the soil from pounding rainfall, which can close natural gaps between soil particles, and loosening soil through root action. This is why forested areas have the highest infiltration rates of any vegetative types.

The top layer of leaf litter that is not decomposed protects the soil from the pounding action of rain, without this the soil can become far less permeable. In chapparal vegetated areas, the hydrophobic oils in the succulent leaves can be spread over the soil surface with fire, creating large areas of hydrophobic soil.

Other conditions that can lower infiltration rates or block them include dry plant litter that resists re-wetting, or frost. If soil is saturated at the time of an intense freezing period, the soil can become a concrete frost on which almost no infiltration would occur.

Over an entire watershed, there are likely to be gaps in the concrete frost or hydrophobic soil where water can infiltrate. Once water has infiltrated the soil it remains in the soil, percolates down to the ground water table, or becomes part of the subsurface run-off process.

Process

The process of infiltration can continue only if there is room available for additional water at the soil surface. The available volume for additional water in the soil depends on the porosity of the soil and the rate at which previously infiltrated water can move away from the surface through the soil. The maximum rate that water can enter a soil in a given condition is the infiltration capacity. If the arrival of the water at the soil surface is less than the infiltration capacity, all of the water will infiltrate. If rainfall intensity at the soil surface occurs at a rate that exceeds the infiltration capacity, ponding begins and is followed by run-off over the ground surface, once depression storage is filled. This run-off is called Horton overland flow. The entire hydrologic system of a watershed is sometimes Analysed using hydrology transport models, mathematical models that consider infiltration, run-off and channel flow to predict river flow rates and stream water quality.

Infiltration in Wastewater Collection

Robert E. Horton suggested that infiltration capacity rapidly declines during the early part of a storm and then tends towards an approximately constant value after a couple of hours for the remainder of the event.

Previously infiltrated water fills the available storage spaces and reduces the capillary forces drawing water into the pores. Clay particles in the soil may swell as they become wet and thereby reduce the size of the pores. In areas where the ground is not protected by a layer of forest litter, raindrops can detach soil particles from the surface and wash fine particles into surface pores where they can impede the infiltration process.

Infiltration in Wastewater Collection

Infiltration is a component of the general mass balance hydrologic budget. There are several ways to estimate the volume and/or the rate of infiltration of water into a soil. Three excellent estimation methods are the Green-Ampt method, SCS method, Horton's method, and Darcy's law.

General Hydrologic Budget

The general hydrologic budget, with all the components, with respect to infiltration F. Given all the other variables and infiltration is the only unknown, simple algebra solves the infiltration question.

$$F = B_I + P - E - T - ET - S - R - I_A - B_O$$

where,

- F is infiltration, which can be measured as a volume or length;
- B_I is the boundary input, which is essentially the output watershed from adjacent, directly connected impervious areas;

- B_O is the boundary output, which is also related to surface run-off, R, depending on where one chooses to define the exit point or points for the boundary output;
- P is precipitation;
- E is evaporation;
- ET is evapotranspiration;
- S is the storage through either retention or detention areas;
- I_A is the initial abstraction, which is the short term surface storage such as puddles or even possibly detention ponds depending on size;
- R is surface run-off.

The only note on this method is one must be wise about which variables to use and which to omit, for doubles can easily be encountered. An easy example of double counting variables is when the evaporation, E, and the transpiration, T, are placed in the equation as well as the evapotranspiration, ET. ET has included in it T as well as a portion of E.

Green-Ampt

Named for two men; Green and Ampt. The Green-Ampt method of infiltration estimation accounts for many variables that other methods, such as Darcy's law, do not. It is a function of the soil suction head, porosity, hydraulic conductivity and time.

$$\int_0^{F(t)} \frac{1-\psi\Delta\theta}{F+\psi\Delta\theta} dF = \int_0^t K dt$$

where,

- ψ is wetting front soil suction head;
- θ is water content;
- K is Hydraulic conductivity;
- F is the total volume already infiltrated.

Once integrated, one can easily choose to solve for either volume of infiltration or instantaneous infiltration rate:

$$F(t) = Kt + \psi\Delta\theta \ln\left[1 + \frac{F(t)}{\psi\Delta\theta}\right].$$

Using this model one can find the volume easily by solving for F(t). However the variable being solved for is in the equation itself so when solving for this one must set the variable in question to converge on zero, or another appropriate constant. A good first guess for F is the larger value of either Kt or $\sqrt{2\psi\Delta\theta Kt}$. The only note on using this formula is that one must assume that h_0, the water head or the depth of ponded water above the surface, is negligible. Using the infiltration volume from this equation one may then substitute F into the corresponding infiltration rate equation below to find the instantaneous infiltration rate at the time, t, F was measured.

$$f(t)=K\left[\frac{\psi\Delta\theta}{F(t)}+1\right].$$

Horton's Equation

Horton's equation is another viable option when measuring ground infiltration rates or volumes. It is an empirical formula that says that infiltration starts at a constant rate, f_0, and is decreasing exponentially with time, t. After some time when the soil saturation level reaches a certain value, the rate of infiltration will level off to the rate f_c.

$$f_t = f_c + (f_0 - f_c)e^{-kt}$$

Where,

- f_t is the infiltration rate at time t;
- f_0 is the initial infiltration rate or maximum infiltration rate;
- f_c is the constant or equilibrium infiltration rate after the soil has been saturated or minimum infiltration rate;
- k is the decay constant specific to the soil.

The other method of using Horton's equation is as below. It can be used to find the total volume of infiltration, F, after time t.

$$F_t = f_c\,t + \frac{(f_0 - f_c)}{k}\left(1 - e^{-kt}\right)$$

Kostiakov Equation

Named after its founder Kostiakov is an empirical equation which assumes that the intake rate declines over time just as to a power function.

$$f(t) = akt^{a-1}$$

where a and k are empirical parameters.

The major limitation of this expression is its reliance on the zero final intake rate. In most cases the infilration rate instead approaches a finite steady value, which in some cases may occur after short periods of time. The Kostiakov-Lewis variant, also known as the "Modified Kostiakov" equation corrects for this by adding a steady intake term to the original equation.

$$f(t) = akt^{a-1} + f_0$$

in integrated form the cumulative volume is expressed as:

$$F(t) = kt^{a} + f_0\,t$$

where f_0 approximates, but does not necessarily equate to the final infiltration rate of the soil.

Darcy's Law

This method used for infiltration is using a simplified version of Darcy's law. In this model the ponded water is assumed to be equal to h0 and the head of dry soil that exists below the depth of the wetting front soil suction head is assumed to be equal to$-\psi-$L.

$$f = k\left[\frac{h_0 - (-\psi - L)}{L}\right]$$

where

h_0 is the depth of ponded water above the ground surface;

K is the hydraulic conductivity;

L is the total depth of subsurface ground in question.

In summary all of these equations should provide a relatively accurate assessment of the infiltration characteristics of the soil in question.

SURFACE RUN-OFF

Surface run-off is the water flow that occurs when soil is infiltrated to full capacity and excess water from rain, meltwater, or other sources flows over the land. This is a major component of the water cycle. Run-off that occurs on surfaces before reaching a channel is also called a Non-point source. If a Non-point source contains man–made contaminants, the run-off is called Non-point source pollution. A land area which produces run-off that drains to a common point is called a watershed. When run-off flows along the ground, it can pick up soil contaminants including, but not limited to petroleum, pesticides, or fertilizers that become discharge or Non-point source pollution.

Generation

Surface run-off can be generated either by rainfall or by the melting of snow, or glaciers. Snow and glacier melt occur only in areas cold enough for these to form permanently. Typically snowmelt will peak in the spring and glacier melt in the summer, leading to pronounced flow maxima in rivers affected by them. The determining factor of the rate of melting of snow or glaciers is both air temperature and the duration of sunlight. In high mountain regions, streams frequently rise on sunny days and fall on cloudy ones for this reason.

In areas where there is no snow, run-off will come from rainfall. However, not all rainfall will produce run-off because storage from soils can absorb light showers. On the extremely ancient soils of Australia and Southern Africa, proteoid roots with their extremely dense networks of root hairs can absorb so much rainwater as to prevent run-off even when substantial amounts of rain fall. In these regions, even on less infertile cracking clay soils, high amounts of rainfall and potential evaporation are needed to generate any surface run-off, leading to specialised adaptations to extremely variable streams.

Infiltration Excess Overland Flow

This occurs when the rate of rainfall on a surface exceeds the rate at which water can infiltrate the ground, and any depression storage has already been filled. This is called infiltration excess overland flow, Hortonian overland flow or unsaturated overland flow. This more commonly occurs in arid and semi-arid regions, where rainfall intensities are high and the soil infiltration capacity is reduced because of surface sealing, or in paved areas. This occurs largely in city areas where pavements prevent water infiltration.

Saturation Excess Overland Flow

When the soil is saturated and the depression storage filled, and rain continues to fall, the rainfall will immediately produce surface run-off. The level of antecedent soil moisture is one factor affecting the time until soil becomes saturated. This run-off is called saturation excess overland flow or saturated overland flow.

Antecedent Soil Moisture

Soil retains a degree of moisture after a rainfall. This residual water moisture affects the soil's infiltration capacity. During the next rainfall event, the infiltration capacity will cause the soil to be saturated at a different rate. The higher the level of antecedent soil moisture, the more quickly the soil becomes saturated. Once the soil is saturated, run-off occurs.

Subsurface Return Flow

After water infiltrates the soil on an up–slope portion of a hill, the water may flow laterally through the soil, and exfiltrate closer to a channel. This is called subsurface return flow or throughflow.

As it flows, the amount of run-off may be reduced in a number of possible ways: a small portion of it may evapotranspire; water may become temporarily stored in microtopographic depressions; and a portion of it may become run-on, which is the infiltration of run-off as it flows overland. Any remaining surface water eventually flows into a receiving water body such as a river, lake, estuary or ocean.

Human Impact on Surface Run-off

Urbanization increases surface run-off, by creating more impervious surfaces such as pavement and buildings, that do not allow percolation of the water down through the soil to the aquifer. It is instead forced directly into streams or storm water run-off drains, where erosion and siltation can be major problems, even when flooding is not. Increased run-off reduces groundwater recharge, thus lowering the water table and making droughts worse, especially for farmers and others who depend on the water wells.

When anthropogenic contaminants are dissolved or suspended in run-off, the human impact is expanded to create water pollution. This pollutant load can reach various receiving waters such as streams, rivers, lakes, estuaries and oceans with resultant water chemistry changes to these water systems and their related ecosystems. A 2008 report by the United States National Research Council identified urban stormwater as a leading source of water quality problems in the U.S.

Effects of Surface Run-off

Erosion and Deposition

Surface run-off causes erosion of the Earth's surface; deposition is the depositing of erosion. There are four principal types of erosion: splash erosion, gully erosion, sheet erosion and stream bed erosion. Splash erosion is the result of mechanical collision of raindrops with the soil surface. Dislodged soil particles becoming suspended in the surface run-off and carried into streams and rivers. Gully erosion occurs when the power of run-off is strong enough that it cuts a well defined channel. These channels can be as small as one centimeter wide or as large as several meters. Sheet erosion is the overland transport of run-off without a well defined channel. In the case of gully erosion, large amounts of material can be transported in a small time period. Stream bed erosion is the attrition of stream banks or bottoms by rapidly flowing rivers or creeks. Reduced crop productivity usually results from erosion, and these effects are studied in the field of soil conservation.

The soil particles carried in run-off vary in size from about.001 millimeter to 1.0 millimeter in diameter. Larger particles settle over short transport distances, whereas small particles can be carried over long distances suspended in the water column. Erosion of silty soils that contain smaller particles generates turbidity and diminishes light transmission, which disrupts aquatic ecosystems.

Entire sections of countries have been rendered unproductive by erosion. On the high central plateau of Madagascar, approximately ten per cent of that country's land area, virtually the entire landscape is devoid of vegetation, with erosive gully furrows typically in excess of 50 meters deep and one kilometer wide. Shifting cultivation is a farming system which sometimes incorporates the slash and burn method in some regions of the world. Erosion cause loss of the fertile top soil and reduces the its fertility and quality of the agricultural produce.

Modern industrial farming is another major cause of erosion. In some areas in the American corn belt, more than 50 per cent of the original topsoil has been carried away within the last 100 years.

Environmental Impacts

The principal environmental issues associated with run-off are the impacts to surface water, groundwater and soil through transport of water pollutants to these systems. Ultimately these consequences translate into human health risk, ecosystem disturbance and aesthetic impact to water resources.

Some of the contaminants that create the greatest impact to surface waters arising from run-off are petroleum substances, herbicides and fertilizers. Quantitative uptake by surface run-off of pesticides and other contaminants has been studied since the 1960s, and early on contact of pesticides with water was known to enhance phytotoxicity.

In the case of surface waters, the impacts translate to water pollution, since the streams and rivers have received run-off carrying various chemicals or sediments. When surface waters are used as potable water supplies, they can be compromised regarding health risks and drinking water aesthetics. Contaminated surface waters risk altering the metabolic processes of the aquatic species that they host; these alterations can lead to death, such as fish kills, or alter the balance of populations present. Other specific impacts are on animal mating, spawning, egg and larvae viability, juvenile survival and plant productivity. Some researches show surface run-off of pesticides, such as DDT, can alter the gender of fish species genetically, which transforms male into female fish.

In the case of groundwater, the main issue is contamination of drinking water, if the aquifer is abstracted for human use. Regarding soil contamination, run-off waters can have two important pathways of concern.

Firstly, run-off water can extract soil contaminants and carry them in the form of water pollution to even more sensitive aquatic habitats. Secondly, run-off can deposit contaminants on pristine soils, creating health or ecological consequences.

Flooding

Flooding occurs when a watercourse is unable to convey the quantity of run-off flowing downstream. The frequency with which this occurs is described by a return period. Flooding is a natural process, which maintains ecosystem composition and processes, but it can also be altered by land use changes such as river engineering. Floods can be both beneficial to societies or cause damage.Agriculture along the Nile floodplain took advantage of the seasonal flooding that deposited nutrients beneficial for crops. However, as the number and susceptibility of settlements increase, flooding increasingly becomes a natural hazard. Adverse impacts span loss of life, property damage, contamination of water supplies, loss of crops, and social dislocation and temporary homelessness. Floods are among the most devastating of natural disasters.

Agricultural Issues

A common context of run-off deals with agriculture. When farmland is tilled and bare soil is revealed, rainwater carries billions of tons of topsoil into waterways each year, causing loss of valuable topsoil and adding sediment to produce turbidity in surface waters.

The other context of agricultural issues involves the transport of agricultural chemicals via surface run-off. This result occurs when chemical use is excessive or poorly timed with respect to high precipitation.

The resulting contaminated run-off represents not only a waste of agricultural chemicals, but also an environmental threat to downstream ecosystems. The alternative to conventional farming is organic farming which eliminates chemical usage.

Measurement and Mathematical Modeling

Run-off is Analysed by using mathematical models in combination with various water quality sampling methods. Measurements can be made using continuous automated water quality analysis instruments targeted on pollutants such as specific organic or inorganic chemicals, pH, turbidity etc. or targeted on secondary indicators such as dissolved oxygen. Measurements can also be made in batch form by extracting a single water sample and conducting any number of chemical or physical tests on that sample.

In the 1950s or earlier hydrology transport models appeared to calculate quantities of run-off, primarily for flood forecasting. Beginning in the early 1970s computer models were developed to Analyse the transport of run-off carrying water pollutants, which considered dissolution rates of various chemicals, infiltration into soils and ultimate pollutant load delivered to receiving waters.

One of the earliest models addressing chemical dissolution in run-off and resulting transport was developed in the early 1970s under contract to the United States Environmental Protection Agency.

This computer model formed the basis of much of the mitigation study that led to strategies for land use and chemical handling controls. Other computer models have been developed that allow surface run-off to be tracked through a river course as reactive water pollutants. In this case the surface run-off may be considered to be a line source of water pollution to the receiving waters.

Mitigation and Treatment

Mitigation of adverse impacts of run-off can take several forms:

- Land use development controls aimed at minimizing impervious surfaces in urban areas
- Erosion controls for farms and construction sites
- Flood control programmes
- Chemical use and handling controls in agriculture, landscape maintenance, industrial use, etc.

Land use Controls

Many world regulatory agencies have encouraged research on methods of minimizing total surface run-off by avoiding unnecessary hardscape. Many municipalities have produced guidelines and codes for land developers that encourage minimum width sidewalks, use of pavers set in earth for driveways and walkways and other design techniques to allow maximum water infiltration in urban settings. An example land use control programme can be viewed at seen in the city of Santa Monica, California.

Erosion Controls

Erosion controls have appeared since medieval times when farmers realised the importance of contour farming to protect soil resources. Beginning in the 1950s these agricultural methods became increasingly more sophisticated. In the 1960s some state and local governments began to focus their efforts on mitigation of construction run-off by requiring builders to implement erosion and sediment controls.

This included such techniques as: use of straw bales and barriers to slow run-off on slopes, installation of silt fences, programming construction for months that have less rainfall and minimizing extent and duration of exposed graded areas. Montgomery County, Maryland implemented the first local government sediment control programme in 1965, and this was followed by a statewide programme in Maryland in 1970.

Flood Control Programmes

Flood control programmes as early as the first half of the twentieth century became quantitative in predicting peak flows of riverine systems. Progressively strategies have been developed to minimize peak flows and also to reduce channel velocities. Some of the techniques commonly applied are: provision of holding ponds to buffer riverine peak flows, use of energy dissipators in channels to reduce stream velocity and land use controls to minimize run-off.

Chemical Use and Handling

Following enactment of the U.S. Resource Conservation and Recovery Act in 1976, and later the Water Quality Act of 1987, states and cities have become more vigilant in controlling the containment and storage of toxic chemicals, thus preventing releases and leakage. Methods commonly applied are: requirements for double containment of underground storage tanks, registration of hazardous materials usage, reduction in numbers of allowed pesticides and more stringent regulation of fertilizers and herbicides in landscape maintenance. In many industrial cases, pretreatment of wastes is required, to minimize escape of pollutants into sanitary or stormwater sewers.

The U.S. Clean Water Act requires that local governments in urbanized areas obtain stormwater discharge permits for their drainage systems. Essentially this means that the locality must operate a stormwater management programme for all surface run-off that enters the municipal separate storm sewer system.

EPA and state regulations and related publications outline six basic components that each local programme must contain:

1. Public education
2. Public involvement
3. Illicit discharge detection and elimination
4. Construction site run-off controls
5. Post–construction storm water management controls
6. Pollution prevention and "good housekeeping" measures.

Other property owners which operate storm drain systems similar to municipalities, such as state highway systems, universities, military bases and prisons, are also subject to the MS4 permit requirements.

SNOWMELT

In hydrology, snowmelt is surface run-off produced from melting snow. It can also be used to describe the period or season during which such run-off is produced. Water produced by snowmelt is an important part of the annual water cycle in many parts of the world, in some cases contributing high fractions of the annual run-off in a watershed. Predicting snowmelt run-off from a drainage basin may be a part of designing water control projects. Rapid snowmelt can cause flooding.

If the snowmelt is then frozen, very dangerous conditions and accidents can occur, introducing the need for salt to melt the ice. Increased water run-off due to snowmelt was a cause of many famous floods. One well-known example is the Red River Flood of 1997, when the Red River of the North in the Red River Valley of the United States and Canada flooded. Flooding in the Red River Valley is augmented by the fact that the river flows north through Winnipeg, Manitoba and into Lake Winnipeg. As snow in Minnesota, North Dakota, and South Dakota begins to melt and flow into the Red River, the presence of downstream ice can act as a dam and force upstream water to rise.

Colder temperatures downstream can also potentially lead to freezing of water as it flows north, thus augmenting the ice dam problem. Some areas in British Columbia are also prone to snowmelt flooding as well.

Energy Fluxes Related to Snowmelt

There are several energy fluxes involved in the melting of snow. These fluxes can act in opposing directions, that is either delivering heat to or removing heat from the snowpack. Ground heat flux is the energy delivered to the snowpack from the soil below by conduction. Radiation inputs to the snowpack include net shortwave and longwave radiation.

Net shortwave radiation is the difference in energy received from the sun and that reflected by the snowpack due to the snowpack albedo. Longwave radiation is received by the snowpack from many sources, including ozone, carbon dioxide, and water vapour present in all levels of the atmosphere. Longwave radiation is also emitted by the snowpack in the form near-Black-body radiation, where snow has an emissivity between 0.97 and 1.0. Generally the net longwave radiation term is negative, meaning a net loss of energy from the snowpack. Latent heat flux is the energy removed from or delivered to the snowpack which accompanies the mass transfers of evaporation, sublimation, or condensation. Sensible heat flux is the heat flux due to convection between the air and snowpack.

INTERCEPTION

Interception, or canopy interception, refers to precipitation that does not reach the soil, but is instead intercepted by the leaves and branches of plants. It occurs in the canopy, and in the forest ground litter. Because of evaporation, interception of liquid water generally leads to loss of that precipitation for the drainage basin, except for cases such as fog interception. Intercepted snowfall does not result in any notable amount of evaporation, and most of the snow falls off the tree by wind or melt.

However, intercepted snow can more easily drift with the wind, out of the watershed. Conifers have greater interception than hardwoods. Their needles gives them more surface area for droplets to adhere to, and they have foliage in spring and fall, therefore interception also depends on the type of vegetation in a wooded area.

RESIDENCE TIMES

The residence time of a reservoir within the hydrologic cycle is the average time a water molecule will spend in that reservoir. It is a measure of the average age of the water in that reservoir.

Groundwater can spend over 10,000 years beneath Earth's surface before leaving. Particularly old groundwater is called fossil water. Water stored in the soil remains there very briefly, because it is spread thinly across the Earth, and is readily lost by evaporation, transpiration, stream flow, or groundwater recharge. After evaporating, the residence time in the atmosphere is about 9 days before condensing and falling to the Earth as precipitation. The major ice sheets–Antarctica and Greenland–store ice for very long periods. Ice from Antarctica has been reliably dated to 800,000 years before present, though the average residence time is shorter.

In hydrology, residence times can be estimated in two ways. The more common method relies on the principle of conservation of mass and assumes the amount of water in a given reservoir is roughly constant. With this method, residence times are estimated by dividing the volume of the reservoir by the rate by which water either enters or exits the reservoir. Conceptually, this is equivalent to timing how long it would take the reservoir to become filled from empty if no water were to leave. An alternative method to estimate residence times, which is gaining in popularity for dating groundwater, is the use of isotopic techniques. This is done in the subfield of isotope hydrology.

CHANGES OVER TIME

The water cycle describes the processes that drive the movement of water throughout the hydrosphere. However, much more water is "in storage" for long periods of time than is actually moving through the cycle. The storehouses for the vast majority of all water on Earth are the oceans. It is estimated that of the 332,500,000 mi^3 of the world's water supply, about 321,000,000 mi^3 is stored in oceans, or about 95%. It is also estimated that the oceans supply about 90% of the evaporated water that goes into the water cycle.

During colder climatic periods more ice caps and glaciers form, and enough of the global water supply accumulates as ice to lessen the amounts in other parts of the water cycle. The reverse is true during warm periods. During the last ice age glaciers covered almost one-third of Earth's land mass, with the result being that the oceans were about 400 ft lower than today. During the last global "warm spell," about 125,000 years ago, the seas were about 18 ft higher than they are now. About three million years ago the oceans could have been up to 165 ft higher. The scientific consensus expressed in the 2007 Intergovernmental Panel on Climate Change Summary for Policymakers is for the water cycle to continue to intensify throughout the 21st century, though this does not mean that precipitation will increase in all regions. In subtropical land areas—places that are already relatively dry—precipitation is projected to decrease during the 21st century, increasing the probability of drought.

The drying is projected to be strongest near the poleward margins of the subtropics. Annual precipitation amounts are expected to increase in near-equatorial regions that tend to be wet in the present climate, and also at high latitudes. These large-scale patterns are present in nearly all of the climate model simulations conducted at several international research centers as part of the 4th Assessment of the IPCC.

Glacial retreat is also an example of a changing water cycle, where the supply of water to glaciers from precipitation cannot keep up with the loss of water from melting and sublimation. Glacial retreat since 1850 has been extensive.

Human activities that alter the water cycle include:

- Agriculture
- Industry
- Alteration of the chemical composition of the atmosphere
- Construction of dams
- Deforestation and afforestation
- Removal of groundwater from wells
- Water abstraction from rivers
- Urbanization

EFFECTS ON CLIMATE

The water cycle is powered from solar energy. 86% of the global evaporation occurs from the oceans, reducing their temperature by evaporative cooling. Without the cooling, the effect of evaporation on the greenhouse effect would lead to a much higher surface temperature of 67 °C and a warmer planet.

Aquifer drawdown or overdrafting and the pumping of fossil water increases the total amount of water in the hydrosphere that is subject to transpiration and evaporation thereby causing accretion in water vapour and cloud cover which are the primary absorbers of infrared radiation in the Earth's atmosphere. Adding water to the system has a forcing effect on the whole earth system, an accurate estimate of which hydrogeological fact is yet to be quantified.

EFFECTS ON BIOGEOCHEMICAL CYCLING

While the water cycle is itself a biogeochemical cycle, flow of water over and beneath the Earth is a key component of the cycling of other biogeochemicals. Run-off is responsible for almost all of the transport of eroded sediment and phosphorus from land to waterbodies. The salinity of the oceans is derived from erosion and transport of dissolved salts from the land. Cultural eutrophication of lakes is primarily due to phosphorus, applied in excess to agricultural fields in fertilizers, and then transported overland and down rivers. Both run-off and groundwater flow play significant roles in transporting nitrogen from the land to waterbodies. The dead zone at the outlet of the Mississippi River is a consequence of nitrates from fertilizer being carried off agricultural fields and funnelled down the river system to the Gulf of Mexico. Run-off also plays a part in the carbon cycle, again through the transport of eroded rock and soil.

SLOW LOSS OVER GEOLOGIC TIME

The hydrodynamic wind within the upper portion of a planet's atmosphere allows light chemical elements such as Hydrogen to move up to the exobase, the lower limit of the exosphere, where the gases can then reach escape velocity, entering outer space without impacting other particles of gas. This type of gas loss from a planet into space is known as planetary wind. Planets with hot lower atmospheres could result in humid upper atmospheres that accelerate the loss of hydrogen.

7

Water Quality Standards and Classifications

Water quality is a term used here to express the suitability of water to sustain various uses or processes. Any particular use will have certain requirements for the physical, chemical or biological characteristics of water; for example limits on the concentrations of toxic substances for drinking water use, or restrictions on temperature and pH ranges for water supporting animals. Consequently, water quality can be defined by a range of variables which limit the water. Although many uses have some common requirements for certain variables, each use will have its own demands and influences on water quality.

Quantity and quality demands of different users will not always be compatible, and the activities of one user may restrict the activities of another, either by demanding water of a specific quality outside the range required by the other user or by lowering quality during use of the water. Efforts to improve or maintain a certain water quality often compromise between the quality and quantity demands of different users. There is an increasing recognition that natural ecosystems have a legitimate place in the consideration of options for water quality management. This is both for their intrinsic value and because they are sensitive indicators of changes or deterioration in overall water quality, providing a useful addition to physical, chemical and other information.

WATER QUALITY

The composition of surface and under groundwater is dependent on natural factors (geological, topographical, meteorological, hydrological and biological) in the drainage basin and varies with seasonal differences in run-off volumes, weather conditions and water levels. Large natural variations in water quality may, therefore, be observed even where only a single watercourse is involved. Human intervention also has significant effects on water quality. Some of these effects are the result of hydrological changes, such as building of dams, draining of diversion of waste waters. More obvious are the polluting activities, such as the discharge of domestic, industrial, urban and other wastewaters into the watercourse (whether intentional or accidental) and the spreading of chemicals on agricultural land in the drainage basin.

Water quality is affected by a wide range of natural and human influences. The most important of the natural influences are geological, hydrological and climatic, since these affect the quantity and the quality of water available.

Their influence is generally higher when available water quantities are low and maximum use is to be made of the limited resource; for example, high salinity is a frequent problem in arid and coastal areas. If the financial and technical resources are available, seawater or saline groundwater can be desalinated but in many circumstances this is not economically feasible. Thus, although water may be available in adequate quantities, its unsuitable quality limits the uses that can be made of it.

Although the natural ecosystem is in harmony with natural water quality, any significant changes to water quality will usually be disruptive to the ecosystem. The effects of human activities on water quality are both widespread and varied in the degree to which they disrupt the ecosystem and/or restrict water use. Pollution of water by human activities, for example, is attributable to only one source, but the reasons for this type of pollution, its impacts on water quality and the necessary remedial or preventive measures are varied. Faecal pollution may occur because there are no community facilities for waste disposal, because collection and treatment facilities are inadequate or improperly operated, or because on-site sanitation facilities (such as latrines) drain directly into aquifers.

The effects of faecal pollution vary appreciably in space and time. In developing countries intestinal disease is the main problem, while organic load and eutrophication may be of greater concern in developed countries (in the rivers into which the sewage or effluent is discharged and in the sea into which the rivers flow or sewage sludge is dumped).

A single influence may, therefore, give rise to a number of water quality problems, just as a problem may have a number of contributing influences. Eutrophication results not only from point sources, such as wastewater discharges with high nutrient loads (principally nitrogen and phosphorus), but also from diffuse sources such as run-off from livestock feedlots or agricultural land fertilized with organic and inorganic fertilizers. Pollution from diffuse sources, such as agricultural run-off, or from numerous small inputs over a wide area, such as faecal pollution from unsewered settlements, is particularly difficult to control.The quality of water may be described in terms of the concentration and state (dissolved or particulate) of some or all of the organic and inorganic material present in the water, together with certain physical characteristics of the water. It is determined by in situ measurements and by examination of water samples on site or in the laboratory. The main elements of water quality monitoring are, therefore, on-site measurements, the collection and analysis of water samples, the study and evaluation of the analytical results, and the reporting of the findings.

The results of analyses performed on a single water sample are only valid for the particular location and time at which that sample was taken. One purpose of a monitoring programme is, therefore, to gather sufficient data (by means of regular or intensive sampling and analysis) to assess spatial and/or temporal variations in water quality.

WATER TEMPERATURE

Water temperature is not only important to fisherman and industries, but also to the growth of fish and algae. A lot of water is used for cooling purposes in power plants that generate electricity. They need cool water to start with, and they generally release warmer water back to the environment. The temperature of the released water can affect downstream habitats. Temperature also can affect the ability of water to hold oxygen as well as the ability of organisms to resist certain pollutants.

pH

The pH is a measure of how acidic/basic a water sample is. The range goes from 0–14, with 7 being neutral. The pH of less than 7 indicates acidity, whereas a pH of greater than 7 indicates a base. The pH is really a measure of the relative amount of free hydrogen and hydroxyl ions in the water.

Water that has more free hydrogen ions is acidic, whereas water that has more free hydroxyl ions is basic. Since pH can be affected by chemicals in the water, pH is an important indicator of water that is changing chemically.

The pH is reported in "logarithmic units," like the Richter scale, which measures earthquakes. Each number represents a 10-fold change in the acidity/basicness of the water. Water with a pH of 5 is ten times more acidic than water having a pH of six.

Pollution can change a water's pH, which in turn can harm animals and plants living in the water. For instance, water coming out of an abandoned coal mine can have a pH of 2, which is very acidic and would definitely affect any fish and other micro organisms. By using the logarithm scale, this mine-drainage water would be several hundred times more acidic than neutral water so stay out of abandoned mines.

SPECIFIC CONDUCTANCE

Specific conductance is a measure of the ability of water to conduct an electrical current. It is highly dependent on the amount of dissolved solids (such as salt) in the water. Pure water, such as distilled water, will have a very low specific conductance, and sea water will have a high specific conductance.

Rainwater often dissolves airborne gasses and airborne dust while it is in the air, and thus often has a higher specific conductance than distilled water. Specific conductance is an important water-quality measurement because it gives a good idea of the amount of dissolved material in the water.

TURBIDITY

Turbidity is the amount of particulate matter that is suspended in water. Turbidity measures the scattering effect that suspended solids have on light: the higher the intensity of scattered light, the higher the turbidity.

Material that causes water to be turbid include:

- Clay
- Silt
- Finely divided organic and inorganic matter
- Soluble coloured organic compounds
- Plankton
- Microscopic organisms

Turbidity makes the water cloudy or opaque. Turbidity is measured by shining a light through the water and is reported in Nephelometric Turbidity Units (NTU). During periods of low flow (base flow), many rivers are a clear green colour, and turbidities are low, usually less than 10 NTU. During a rainstorm, particles from the surrounding land are washed into the river making the water a muddy brown colour, indicating water that has higher turbidity values. Also, during high flows, water velocities are faster and water volumes are higher, which can more easily stir up and suspend material from the stream bed, causing higher turbidities.

Turbidity can be measured in the laboratory and also on-site in the river. A handheld turbidity meter measures turbidity of a water sample.

The meter is calibrated using standard samples from the meter manufacturer. The three glass vials shows turbidity standards of 5, 50, and 500 NTUs. Once the meter is calibrated to correctly read these standards, the turbidity of a water sample can be taken.

DISSOLVED OXYGEN

Although water molecules contain oxygen atom, this oxygen is not what is needed by aquatic organisms living in natural waters. A small amount of oxygen, upto about ten molecules of oxygen per million of water, is actually dissolved in water. This dissolved oxygen is breathed by fish and zooplankton and is needed by them to survive. Rapidly moving water, such as in a mountain stream or large river, tends to contain a lot of dissolved oxygen, while stagnant water contains little. Bacteria in water can consume oxygen as organic matter decays. Thus, excess organic material in the tanks, lakes and rivers can cause an oxygen-deficient situation to occur. Aquatic life can have a hard time in stagnant water that has a lot of rotting, organic material in it, especially in summer, when dissolved-oxygen levels are at a seasonal low.

HARDNESS

The amount of dissolved calcium and magnesium in water determines its 'hardness.' If some one live in an area where the water is 'soft,' then you may never have even heard of water hardness. But, if one live, where the water is relatively hard, may notice that it is difficult to get lather up when washing your hands or clothes. And, industries in these area might have to spend money to soften their water, as hard water can damage equipment. Hard water can even shorten the life of fabrics and clothes.

SUSPENDED SEDIMENT

Suspended sediment is the amount of soil moving along in a stream. It is highly dependent on the speed of the water flow, as fast-flowing water can pick up and suspend more soil than calm water. During storms, soil is washed from the stream banks into the stream. The amount that washes into a stream depends on the type of land in the river's watershed and the vegetation surrounding the river.

If land is disturbed along a stream and protection measures are not taken, then excess sediment can harm the water quality of a stream. These bunds and fences are supposed to trap sediment during a rainstorm and keep it from washing into a stream, as excess sediment can harm the creeks, rivers, lakes, and reservoirs.

Sediment coming into a reservoir is always a concern; once it enters it cannot get out–most of it will settle to the bottom. Reservoirs can 'silt in' if too much sediment enters them. The volume of the reservoir is reduced, resulting in less area for water storage for agriculture as well as reducing the power-generation capability of the power plant in the dam.

POTABLE WATER QUALITY

Potable water must be free of anything that would degrade human performance. Further, it should not damage the materials used in its transportation and storage. Potable water must be suitable for maintaining human health (personal hygiene and medical treatment). Water quality standards give a basis for selecting or rejecting water intended for human use. These standards provide minimum accepted values for safeguarding human health.

IMPURITIES IN WATER

As water goes through the hydrologic cycle, it gathers many impurities. Dust, smoke, and gases fill the air and can contaminate rain, snow, hail, and sleet. The rain water picks up silt, chemicals, and disease organisms. As it enters the earth through seepage and infiltration, some of the suspended impurities may be filtered out. However, other minerals and chemicals are dissolved and carried along. As groundwater it may contain disease organisms as well as harmful chemicals.

In addition to the impurities in water resulting from infiltration, many are contributed by an industrialized society. The garbage, sewage, industrial waste, pesticides etc., are all possible contaminants of raw water. Impurities in raw water are either suspended or dissolved. Suspended impurities include diseases organisms, silt, bacteria, and algae.They must be removed or destroyed before the water is consumed. Dissolved impurities include salts, (calcium, magnesium, and sodium), iron, manganese, and gases (oxygen, carbon dioxide, hydrogen sulphide, pH and nitrogen). These impurities must be reduced to levels acceptable for human consumption.

RAW WATER CLASSIFICATIONS

Water is classified as fresh, brackish, or salt water (seawater) based on the concentration of TDS (Total Dissolved Solids). Fresh water has a TDS concentration of less that 1,500 ppm. Brackish water is high in minerals and has a TDS concentration between 1,500 ppm and 16,000 ppm. Salt water has a TDS concentration greater than 15,000 ppm.

Generally, groundwater (subsurface) has less chemical or biological contaminants than surface water, provided reasonable care is exercised in the selection of the well site. Harmful micro organisms are usually reduced to tolerable levels by passage through the soil.

TREATED WATER CLASSIFICATIONS

- *Palatable water*: Palatable water is water that is pleasing in appearance and taste. It is significantly free from colour, turbidity, taste, and odour. It should also be cool and aerated.
- *Potable water quality standards*: Quality standards for treated water reflect the values of substances allowed in potable water. Standards exist to measure the physical, chemical, microbiologic, and radiologic quality of water and to test for the presence of chemical agents.

Physical Quality

The principal physical characteristics of water are colour, odour and taste, turbidity, and temperature.

These characteristics and their related quality standards are described below:

- *Colour*: Colour in water is derived from coloured substances, such as vegetable matter, dissolved from roots and leaves, from humus, or from inorganic compounds such as iron and manganese salts. The colour standard is designed to make drinking water more palatable.
- *Odour and Taste*: There is no set standards for odour and taste as there are no specific tests for these. Odour and taste found in water are most commonly caused by algae, decomposed organic matter, dissolved gases, or industrial waste.
- *Turbidity*: Turbidity refers to a muddy or unclear condition of water caused by suspended clay, silt, organic and inorganic matter, and plankton and other microorganisms. The turbidity standard was established to improve the efficiency of disinfection by reducing particles to which micro organisms could attach.
- *Temperature:* Warm water tastes flat. Cooling water suppresses odours and tastes and makes it more palatable. Temperature also effects the chlorination and purification of water. Disinfection takes longer when water is colder, and purification capacity is reduced with reverse osmosis treatment equipment. Water having physical characteristics exceeding the limits or making it less palatable should not be used for drinking. Otherwise, reduced consumption and increased risk of dehydration may result. When water of low physical quality has not been used, the appropriate command level will make that decision based on medical recommendations.

CHEMICAL QUALITY

The chemical quality of water depends on the chemical substances it contains. These substances include TDS, chlorides, sulphates and other ions.

The chemical quality of water involves its hardness, alkalinity, acidity, and corrosiveness. Chemical substances having an adverse health effect have established standards that will not be exceeded without medical approval.

Potential Hydrogen

The pH is a measure of the acidic or alkaline nature of water. It is technically defined as the negative logarithm of the hydrogen ion concentration. It ranges from 0 to 14 and pH of 7 is neutral.

The pH influences the corrosiveness of the water, the amount of chemicals needed for proper disinfection, and the ability of an analyst to detect contaminants. Water with a pH below 7 is regarded as acidic while that with a pH above 7 is regarded as alkaline. The pH standard was established to ensure effective purification and disinfection.

Arsenic

Arsenic can be present in natural water sources in a wide range of concentrations. It can come from either natural or industrial sources. Ingestion of low concentrations of arsenic can cause nausea, vomiting, abdominal pain, or nerve damage.

Chloride

Chloride exists in most natural waters. It is the main anion found in seawater. Chloride comes from natural salt deposits, domestic and industrial waste, and agricultural run-off. Even in low concentrations, chloride can produce an objectionable taste in water. The chloride standard ensures that potable water is also palatable.

Calcium, Magnesium and Hardness

Calcium and magnesium in drinking water, is an important parameter is to be taken into account: water hardness, even if this term is incorrect and obsolete from a strictly chemical point of view. That is to say that both of these elements largely have not been analysed individually in drinking water in the past, but just non-specifically in summary as hardness.

This approach was applied in many studies focused on health effects of this 'water factor'. Since the definition of water hardness is approached either analytically or technologically, it was not and still has not been defined in a unified manner, and as with other parameters, multiple definitions have been available and multiple units have been used to express it (German, French, and English degrees; equivalent $CaCO_3$ or CaO in mg/l). Initially, water hardness was understood to be a measure of the capacity of water to precipitate soap, which is in practice the sum of concentrations of all polyvalent cations present in water (Ca, Mg, Sr, Ba, Fe, Al, Mn, etc.); nevertheless, since the other ions (apart from Ca and Mg) play a minor role in this regard, later it has been generally accepted that hardness is defined as the sum of the Ca and Mg concentrations.

From the technical point of view, multiple different scales of water hardness were suggested (*i.e.*, very soft–soft–medium hard–hard–very hard).

Expectedly, both extreme degrees (*i.e.* very soft and very hard) are considered as undesirable concordantly from the technical and health points of view, but the optimum Ca and Mg water levels are not easy to determine since the health requirements may not coincide with the technical ones.

Calcium and Magnesium Presence in Waters

Water calcium and magnesium result from decomposition of calcium and magnesium aluminosilicates and, at higher concentrations, from dissolution of limestone, magnesium limestone, magnesite, gypsum and other minerals. Anthropogenenic contamination of drinking water sources with calcium and magnesium is not common but drinking water may be intentionally supplemented with these elements while treated, as happens with deacidification of underground waters by means of calcium hydroxide or filtration through different compounds counteracting acidity such as $CaCO_3$, $MgCO_3$ and MgO, and possibly also with stabilization of low-mineralized waters by addition of CaO and CO_2.

In low-and medium-mineralized underground and surface waters (as drinking waters are), calcium and magnesium are mainly present as simple ions Ca^{2+} and Mg^{2+}, the Ca levels varying from tens to hundreds of mg/l and the Mg concentrations varying from units to tens of mg/l.

Magnesium is usually less abundant in waters than calcium, which is easy to understand since magnesium is found in the Earth's crust in much lower amounts as compared with calcium. In common underground and surface waters the weight concentration of Ca is usually several times higher compared to that of Mg, the Ca to Mg ratio reaching up to 10. Nevertheless, a common Ca to Mg ratio is about 4, which corresponds to a substance ratio of 2.4.

Sulphates

Sulphates occur naturally in water as the result of dissolution of sulfurbearing minerals. Significant concentrations also result from industry sources, such as coal mine drainage, pulp paper mills, tanneries, textile mills, and domestic waste water. When ingested, sulphates have a laxative effect. They also can produce a bad taste in water. The sulphate standard was established to prevent chemically induced diarrhea.

Total Dissolved Solids

The TDS of water is composed of mineral salts and small amounts of other inorganic and organic substances. The proportion of each constituent is the result of weathering of rocks found in the drainage basin and of any industrial contributions. Since TDS is composed of chloride, magnesium, sulfate, and other ions, its ingestion in water has the same effects. Therefore, the TDS standard was established to prevent chemically induced diarrhea.

Chemical water quality standards are based on the effect the water will have on the health of living organisms. The effect of a particular chemical substance determines if a limit is established for that substance. Chemical substances having a negative physical effect will have a mandatory limit that should not be exceeded. Some substances, such as iron and manganese, have no significant negative physical effect, but may restrict the use of the water, such as for the laundering of clothes.

MICROBIOLOGICAL QUALITY

The microbiological quality of potable water shows its potential for transmitting waterborne diseases. These diseases may be caused by viruses, bacteria, protozoa, or higher organisms. A microbiological test will reveal the quality of the raw water source and aid in determining any treatment required.The test is necessary to maintain the quality of the water. The testing for microorganisms in water is extremely difficult. The number of these organisms is usually very low, even in a badly polluted water supply, and the test used to find them is difficult. For these reasons, indicator organisms are used to detect the presence of contamination.

The bacterial organisms used as an indicator of possible contamination are total coliform. These organisms occur in large quantities in the intestines of warm-blooded animals.

The presence of any coliform organism in treated potable water is an indication of either inadequate treatment or the introduction of undesirable materials to the water after treatment.

While the detection of many disease-causing microbes is difficult, the test to detect a surrogate organism, E. coli, is simple and effective for field use.

Because of its relative simplicity and field adaptability, the membrane filter technique has gained wide acceptance as the preferred technique for the presumptive determination of the presence of coliform organisms in potable water. The microbiological standard was established to ensure infectious microorganisms would not cause diseases.

WATER QUALITY STANDARDS

The idea of water quality management is to ensure that water supplied is free from pathogenic organisms, clear, potable and free from undesirable taste and odour, of reasonable temperature, either corrosive nor scale forming and free from minerals which could produce undesirable physiological effect.

The establishment of minimum standards of quality for public water supply is of fundamental importance in achieving this ideal. Standards of quality from the yardstick with which the quality control of any public water supply has to be assessed. In India, certain minimum standards have already been prescribed and given here.

STANDARDS FOR POTABLE AND SAFE WATER

National Drinking Water Mission is an institution which is the in-charge to define the standards for potable water with respects to their locations. It is defined as the water that is free from pathogenic micro-organisms, poisonous substances, excessive amounts of minerals and organic matter which would produce undesirable physiological effects. It should be free from colour, turbidity, taste and odour, of moderate temperature and aerated.

The physical and chemical quality of water should not excess the limits shown in the table below:

Table. Physical and Chemical Standards

Sl. No.	**Characteristics**	*** Acceptable**	**** Cause for Rejection**	
1.	Turbidity (Units on JTU scale)		2.5	10
2.	Colour (units on platinum-cobalt scale)		5.0	25
3.	Taste and odour	Unobjectionable	Unobjectionable	
4.	pH	7.0 to 8.5	6.5 to 9.2	
5.	Total dissolved solids	500	1500	
6.	Total hardness (mg/l as $CaCO_3$)		200	600
7.	Chlorides (mg/l as Cl)	200	1000	
8.	Sulphates (as SO_4, mg/l)		200	1000
9.	Fluorides (as F, mg/l)	1.0	1.5	
10.	Nitrates (as NO_3, mg/l)		45	100
11.	Calcium (as Ca, mg/l)	75	200	
12.	Magnesium (as Mg, mg/l)		30	150
13.	Iron (as Fe, mg/l)	0.1	1.0	
14.	Manganese (as Mn, mg/l)		0.05	0.5

Notes:

* The figure indicated under the column acceptable are the limits upto which the water is generally acceptable to the customers.

** Figures in excess of those mentioned under acceptable render the water non-acceptable but still may be tolerated in the absence of alternative and better source but upto the limits indicated under column cause for rejection above which the supply will have to be rejected.

Bacteriological Standards

- Coliform count in any sample of 100 ml should be zero. A sample of water entering the distribution system that does not conform to this standard calls an immediate investigation into both efficacy of the purification and the method of sampling.
- *Water in the distribution system shall satisfy all the three criteria indicated below*:
  - E,Coli count in 100 ml of any sample should be zero
  - Coliform organisms not more that 10 per 100 ml shall be present in any sample.

– Coliform organisms should not be detectable in 100 ml of any two consecutive samples or more than 50% of the samples collected for the year

The World Health Organization (WHO) is a specialized agency of the United Nations (UN) that acts as a coordinating authority on international public health. Established on 7 April 1948, and headquartered in Geneva, Switzerland, the agency inherited the mandate and resources of its predecessor, the Health Organization, which had been an agency of the League of Nations. Guideline Values for chemicals that are health significance in drinking–water.

Table. WHO Guidelines (Values for Health Related Organic Contaminants)

Chemicals	Guideline Values (mg/litre)
Arsenic	0.01
Fluoride	1.5
Manganese	0.4
Nitrate	50
Nitrite	3
Chlorine	5
Copper	2000
Lead	10
Nickel	20

These Guidelines provide a generally applicable approach to drinking–water safety. Their application to drinking-water supply through piped distribution and through community supplies are described. In applying the Guidelines in specific circumstances, additional factors are also important.

8

Methods of Water Treatment

Water treatment describes those processes used to make water more acceptable for a desired end-use. These can include use as drinking water, industrial processes, medical and many other uses. The goal of all water treatment process is to remove existing contaminants in the water, or reduce the concentration of such contaminants so the water becomes fit for its desired end-use. One such use is returning water that has been used back into the natural environment without adverse ecological impact.

The processes involved in treating water for drinking purpose may be solids separation using physical such as settling and filtration, chemical such as disinfection and coagulation. Biological processes are also employed in the treatment of wastewater and these processes may include, for example, aerated lagoons, activated sludge or slow sand filters.

WATER PURIFICATION

Water purification is the process of removing undesirable chemicals, materials, and biological contaminants from raw water. The goal is to produce water fit for a specific purpose. Most water is purified for human consumption but water purification may also be designed for a variety of other purposes, including to meet the requirements of medical, pharmacology, chemical and industrial applications.

In general the methods used include physical process such as filtration and sedimentation, biological processes such as slow sand filters or activated sludge, chemical process such as flocculation and chlorination and the use of electromagnetic radiation such as ultraviolet light. The purification process of water may reduce the concentration of particulate matter including suspended particles, parasites, bacteria, algae, viruses, fungi; and a range of dissolved and particulate material derived from the surfaces that water may have made contact with after falling as rain. The standards for drinking water quality are typically set by governments or by international standards. These standards will typically set minimum and maximum concentrations of contaminants for the use that is to be made of the water. It is not possible to tell whether water is of an appropriate quality by visual examination. Simple procedures such as boiling or the use of a household activated carbon filter are not sufficient for treating all the possible contaminants that may be present in water from an unknown source. Even natural spring water - considered safe for all practical purposes in the 1800s - must now be tested before determining what kind of treatment, if any, is needed. Chemical analysis, while expensive, is the only way to obtain the information necessary for deciding on the appropriate method of purification. According to a 2007 World Health Organization report, 1.1 billion people lack access to an improved drinking water supply, 88% of the 4 billion annual cases of diarrheal disease are attributed to unsafe water and inadequate sanitation and hygiene, and 1.8 million people die from diarrheal diseases each year.

The WHO estimates that 94% of these diarrheal cases are preventable through modifications to the environment, including access to safe water. Simple techniques for treating water at home, such as chlorination, filters, and solar disinfection, and storing it in safe containers could save a huge number of lives each year.

SOURCES OF WATER

- *Groundwater*: The water emerging from some deep ground water may have fallen as rain many decades, hundreds, thousands or in some cases millions of years ago. Soil and rock layers naturally filter the ground water to a high degree of clarity before it is pumped to the treatment plant. Such water may emerge as springs, artesian springs, or may be extracted from boreholes or wells.

Deep ground water is generally of very high bacteriological quality but the water typically is rich in dissolved solids, especially carbonates and sulfates of calcium and magnesium. Depending on the strata through which the water has flowed, other ions may also be present including chloride, and bicarbonate. There may be a requirement to reduce the iron or manganese content of this water to make it pleasant for drinking, cooking, and laundry use. Disinfection may also be required. Where groundwater recharge is practised; a process in which river water is injected into an aquifer to store the water in times of plenty so that it is available in times of drought; it is equivalent to lowland surface waters for treatment purposes.

- *Upland lakes and reservoirs*: Typically located in the headwaters of river systems, upland reservoirs are usually sited above any human habitation and may be surrounded by a protective zone to restrict the opportunities for contamination. Bacteria and pathogen levels are usually low, but some bacteria, protozoa or algae will be present. Where uplands are forested or peaty, humic acids can colour the water. Many upland sources have low pH which require adjustment.

- *Rivers, canals and low land reservoirs*: Low land surface waters will have a significant bacterial load and may also contain algae, suspended solids and a variety of dissolved constituents.
- Atmospheric water generation is a new technology that can provide high quality drinking water by extracting water from the air by cooling the air and thus condensing water vapor.
- Rainwater harvesting or fog collection which collects water from the atmosphere can be used especially in areas with significant dry seasons and in areas which experience fog even when there is little rain.
- Desalination of seawater by distillation or reverse osmosis.

Treatment

The processes below are the ones commonly used in water purification plants. Some or most may not be used depending on the scale of the plant and quality of the water.

Pre-treatment

- *Pumping and containment*: The majority of water must be pumped from its source or directed into pipes or holding tanks. To avoid adding contaminants to the water, this physical infrastructure must be made from appropriate materials and constructed so that accidental contamination does not occur.
- *Screening*: The first step in purifying surface water is to remove large debris such as sticks, leaves, trash and other large particles which may interfere with subsequent purification steps. Most deep groundwater does not need screening before other purification steps.
- *Storage*: Water from rivers may also be stored in bankside reservoirs for periods between a few days and many months to allow natural biological purification to take place. This is especially important if treatment is by slow sand filters. Storage reservoirs also provide a buffer against short periods of drought or to allow water supply to be maintained during transitory pollution incidents in the source river.
- Pre-conditioning: Many waters rich in hardness salts are treated with soda-ash to precipitate calcium carbonate out utilising the common-ion effect.
- *Pre-chlorination*: In many plants the incoming water was chlorinated to minimise the growth of fouling organisms on the pipe-work and tanks. Because of the potential adverse quality effects, this has largely been discontinued.

Widely varied techniques are available to remove the fine solids, micro-organisms and some dissolved inorganic and organic materials. The choice of method will depend on the quality of the water being treated, the cost of the treatment process and the quality standards expected of the processed water.

pH Adjustment

Distilled water has a pH of 7 and sea water has an average pH of 8.3 . If the water is acidic , lime, soda ash, or sodium hydroxide is added to raise the pH. For somewhat acidic, alkaline waters , forced draft degassifiers are the cheapest way to lower the pH, as the process raises the pH by stripping dissolved carbon dioxide from the water.

Lime is commonly used for pH adjustment for municipal water, or at the start of a treatment plant for process water, as it is cheap, but it also increases the ionic load by raising the water hardness. Making the water slightly alkaline ensures that coagulation and flocculation processes work effectively and also helps to minimize the risk of lead being dissolved from lead pipes and lead solder in pipe fittings.

Acid may be added to alkaline waters in some circumstances to lower the pH. Having an alkaline water does not necessarily mean that lead or copper from the plumbing system will not be dissolved into the water but as a generality, water with a pH above 7 is much less likely to dissolve heavy metals than a water with a pH below 7.

FLOCCULATION

Flocculation is a process which clarifies the water. Clarifying means removing any turbidity or colour so that the water is clear and colourless. Clarification is done by causing a precipitate to form in the water which can be removed using simple physical methods. Initially the precipitate forms as very small particles but as the water is gently stirred, these particles stick together to form bigger particles - this process is sometimes called flocculation. Many of the small particles that were originally present in the raw water absorb onto the surface of these small precipitate particles and so get incorporated into the larger particles that coagulation produces.

In this way the coagulated precipitate takes most of the suspended matter out of the water and is then filtered off, generally by passing the mixture through a coarse sand filter or sometimes through a mixture of sand and granulated anthracite .

Coagulants / flocculating agents that may be used include:

- *Iron (III) hydroxide*: This is formed by adding a solution of an iron (III) compound such as iron(III) chloride to pre-treated water with a pH of 7 or greater. Iron (III) hydroxide is extremely insoluble and forms even at a pH as low as 7. Commercial formulations of iron salts were traditionally marketed in the UK under the name Cuprus.

- Aluminium hydroxide is also widely used as the flocculating precipitate although there have been concerns about possible health impacts and mis-handling led to a severe poisoning incident in 1988 at Camelford in south-west UK when the coagulant was introduced directly into the holding reservoir of final treated water.
- PolyDADMAC is an artificially produced polymer and is one of a class of synthetic polymers that are now widely used. These polymers have a high molecular weight and form very stable and readily removed flocs, but tend to be more expensive in use compared to inorganic materials.

SEDIMENTATION

Water exiting the flocculation basin may enter the sedimentation basin, also called a clarifier or settling basin. It is a large tank with slow flow, allowing floc to settle to the bottom. The sedimentation basin is best located close to the flocculation basin so the transit between does not permit settlement or floc break up.

Sedimentation basins can be in the shape of a rectangle, where water flows from end to end, or circular where flow is from the centre outward. Sedimentation basin outflow is typically over a weir so only a thin top layer - furthest from the sediment - exits.

The amount of floc that settles out of the water is dependent on the time the water spends in the basin and the depth of the basin. The retention time of the water must therefore be balanced against the cost of a larger basin. The minimum clarifier retention time is normally 4 hours. A deep basin will allow more floc to settle out than a shallow basin. This is because large particles settle faster than smaller ones, so large particles bump into and integrate smaller particles as they settle. In effect, large particles sweep vertically through the basin and clean out smaller particles on their way to the bottom.

As particles settle to the bottom of the basin, a layer of sludge is formed on the floor of the tank. This layer of sludge must be removed and treated. The amount of sludge that is generated is significant, often 3%-5% of the total volume of water that is treated.

The cost of treating and disposing of the sludge can be a significant part of the operating cost of a water treatment plant. The tank may be equipped with mechanical cleaning devices that continually clean the bottom of the tank or the tank can be taken out of service when the bottom needs to be cleaned.

FILTRATION

After separating most floc, the water is filtered as the final step to remove remaining suspended particles and unsettled floc. The most common type of filter is a rapid sand filter. Water moves vertically through sand which often has a layer of activated carbon or anthracite coal above the sand. The top layer removes organic compounds, which contribute to taste and Odour. The space between sand particles is larger than the smallest suspended particles, so simple filtration is not enough. Most particles pass through surface layers but are trapped in pore spaces or adhere to sand particles. Effective filtration extends into the depth of the filter. This property of the filter is key to its operation: if the top layer of sand were to block all the particles, the filter would quickly clog.

To clean the filter, water is passed quickly upward through the filter, opposite the normal direction to remove embedded particles. Prior to this, compressed air may be blown up through the bottom of the filter to break up the compacted filter media to aid the backwashing process; this is known as *air scouring*. This contaminated water can be disposed of, along with the sludge from the sedimentation basin, or it can be recycled by mixing with the raw water entering the plant. Some water treatment plants employ pressure filters. These work on the same principle as rapid gravity filters, differing in that the filter medium is enclosed in a steel vessel and the water is forced through it under pressure.

Advantages:

- Filters out much smaller particles than paper and sand filters can.
- Filters out virtually all particles larger than their specified pore sizes.
- They are quite thin and so liquids flow through them fairly rapidly.
- They are reasonably strong and so can withstand pressure differences across them of typically 2-5 atmospheres.
- They can be cleaned and reused.

Membrane filters are widely used for filtering both drinking water and sewage . For drinking water, membrane filters can remove virtually all particles larger than 0.2 um—including Giardia and cryptosporidium. Membrane filters are an effective form of tertiary treatment when it is desired to reuse the water for industry, for limited domestic purposes, or before discharging the water into a river that is used by towns further downstream. They are widely used in industry, particularly for beverage preparation . However no filtration can remove substances that are actually dissolved in the water such as phosphorus, nitrates and heavy metal ions.

SLOW SAND FILTERS

Slow sand filters may be used where there is sufficient land and space as the water must be passed very slowly through the filters. These filters rely on biological treatment processes for their action rather than physical filtration. The filters are carefully constructed using graded layers of sand with the coarsest sand, along with some gravel, at the bottom and finest sand at the top. Drains at the base convey treated water away for disinfection.

Filtration depends on the development of a thin biological layer, called the zoogleal layer or Schmutzdecke, on the surface of the filter. An effective slow sand filter may remain in service for many weeks or even months if the pre-treatment is well designed and produces water with a very low available nutrient level which physical methods of treatment rarely achieve. Very low nutrient levels allow water to be safely sent through distribution system with very low disinfectant levels thereby reducing consumer irritation over offensive levels of chlorine and chlorine by-products. Slow sand filters are not backwashed; they are maintained by having the top layer of sand scraped off when flow is eventually obstructed by biological growth.

A specific 'large-scale' form of slow sand filter is the process of bank filtration, in which natural sediments in a riverbank are used to provide a first stage of contaminant filtration. While typically not sufficiently clean enough to be used directly for drinking water, the water gained from the associated extraction wells is much less problematic than river water taken directly from the major streams where bank filtration is often used.

LAVA FILTERS

Lava filters are similar to sand filters and may also only be used where there is sufficient land and space. Like sand filters, the filters rely on biological treatment processes for their action rather than physical filtration. Unlike slow sand filters however, they are constructed out of 2 layers of lava pebbles and a top layer of nutrient-free soil . On top, water-purifying plants are placed. Usually, around 1/4 of the dimension of lavastone is required to purify the water and just like slow sand filters, a series of herringbone drains are placed .

Removal of Ions and Other Dissolved Substances

Ultrafiltration membranes use polymer membranes with chemically formed microscopic pores that can be used to filter out dissolved substances avoiding the use of coagulants. The type of membrane media determines how much pressure is needed to drive the water through and what sizes of micro-organisms can be filtered out.

Ion exchange: Ion exchange systems use ion exchange resin- or zeolite-packed columns to replace unwanted ions. The most common case is water softening consisting of removal of Ca and Mg ions replacing them with benign

Na or K ions. Ion exchange resins also used to remove toxic ions such as nitrate, nitrite, lead, mercury, arsenic and many others.

Electrodeionization: Water is passed between a positive electrode and a negative electrode. Ion exchange membranes allow only positive ions to migrate from the treated water Towards the negative electrode and only negative ions Towards the positive electrode. High purity deionized water is produced with a little worse degree of purification in comparison with ion exchange treatment. Complete removal of ions from water is regarded as electrodialysis. The water is often pre-treated with a reverse osmosis unit to remove non-ionic organic contaminants.

OTHER MECHANICAL AND BIOLOGICAL TECHNIQUES

In addition to the many techniques used in large-scale water treatment, several small-scale, less polluting techniques are also being used to treat polluted water. These techniques include those based on mechanical and biological processes.

An overview:

- *Mechanical systems*: Sand filtration, lava filter systems and systems based on UV-radiation)
- *Biological systems*: Plant systems as constructed wetlands and treatment ponds and compact systems as activated sludge systems, biorotors, aerobic and anaerobic biofilters, submerged aerated filters, and biorolls

In order to purify the water adequately, several of these systems are usually combined to work as a whole. Combination of the systems is done in two to three stages, namely primary and secondary purification. Sometimes tertiary purification is also added.

Disinfection

Disinfection is accomplished both by filtering out harmful microbes and also by adding disinfectant chemicals in the last step in purifying drinking water. Water is disinfected to kill any pathogens which pass through the filters. Possible pathogens include viruses, bacteria, including *Escherichia coli*, *Campylobacter* and *Shigella*, and protozoans, including *Giardia lamblia* and other cryptosporidia. In most developed countries, public water supplies are required to maintain a residual disinfecting agent throughout the distribution system, in which water may remain for days before reaching the consumer. Following the introduction of any chemical disinfecting agent, the water is usually held in temporary storage - often called a contact tank or clear well to allow the disinfecting action to complete.

- *Chlorination*: The most common disinfection method is some form of chlorine or its compounds such as chloramine or chlorine dioxide.

Chlorine is a strong oxidant that rapidly kills many harmful micro-organisms. Because chlorine is a toxic gas, there is a danger of a release associated with its use. This problem is avoided by the use of sodium hypochlorite, which is a relatively inexpensive solution that releases free chlorine when dissolved in water. Chlorine solutions can be generated on site by electrolyzing common salt solutions. A solid form, calcium hypochlorite exists that releases chlorine on contact with water. Handling the solid, however, requires greater routine human contact through opening bags and pouring than the use of gas cylinders or bleach which are more easily automated. The generation of liquid sodium hypochlorite is both inexpensive and safer than the use of gas or solid chlorine. All forms of chlorine are widely used despite their respective drawbacks. One drawback is that chlorine from any source reacts with natural organic compounds in the water to form potentially harmful chemical by-products trihalomethanes and haloacetic acids, both of which are carcinogenic in large quantities and regulated by the United States Environmental Protection Agency . The formation of THMs and haloacetic acids may be minimized by effective removal of as many organics from the water as possible prior to chlorine addition. Although chlorine is effective in killing bacteria, it has limited effectiveness against protozoans that form cysts in water .

- Chlorine dioxide is another faster-acting disinfectant. It is, however, relatively rarely used, because in some circumstances it may create excessive amounts of chlorite, which is a by-product regulated to low allowable levels in the United States. Chlorine dioxide is made in water and added/used in water to avoid gas handling problems; chlorine dioxide gas accumulations may spontaneously detonate.
- Chloramines are another chlorine-based disinfectant. Although chloramine is not as strong of an oxidant it does provide a longer-lasting residual than free chlorine, and it won't form THMs or haloacetic acids. It is possible to convert chlorine to chloramine by adding ammonia to the water after addition of chlorine: The chlorine and ammonia react to form chloramine. Water distribution systems disinfected with chloramines may experience nitrification, wherein ammonia is used a nutrient for bacterial growth, with nitrates being generated as a byproduct.
- Ozone is an unstable molecule, a "free radical" of oxygen which readily gives up one atom of oxygen providing a powerful oxidising agent which is toxic to most waterborne organisms. It is a very strong, broad spectrum disinfectant that is widely used in Europe.

It is an effective method to inactivate harmful protozoans that form cysts. It also works well against almost all other pathogens. Ozone is made by passing oxygen through ultraviolet light or a "cold" electrical discharge. To use ozone as a disinfectant, it must be created on-site and added to the water by bubble contact. Some of the advantages of ozone include the production of fewer dangerous by-products and the lack of taste and odour produced by ozonation. Although fewer by-products are formed by ozonation, it has been discovered that the use of ozone produces a small amount of the suspected carcinogen bromate, although little bromine should be present in treated water. Another of the main disadvantages of ozone is that it leaves no disinfectant residual in the water. Ozone has been used in drinking water plants since 1906 where the first industrial ozonation plant was built in Nice, France. The U.S. Food and Drug Administration has accepted ozone as being safe; and it is applied as an anti-microbiological agent for the treatment, storage, and processing of foods.

- UV radiation is very effective at inactivating cysts, as long as the water has a low level of colour so the UV can pass through without being absorbed. The main disadvantage to the use of UV radiation is that, like ozone treatment, it leaves no residual disinfectant in the water. Because neither ozone nor UV radiation leaves a residual disinfectant in the water, it is sometimes necessary to add a residual disinfectant after they are used. This is often done through the addition of chloramines, discussed above as a primary disinfectant. When used in this manner, chloramines provide an effective residual disinfectant with very little of the negative aspects of chlorination.
- Hydrogen peroxide works in a similar way to ozone. Activators such as formic acid are often added to increase the efficacy of disinfection. It has the disadvantages that it is slow-working, phytotoxic in high dosage, and decreases the pH of the water it purifies.
- Various portable methods are available for disinfection in emergencies or in remote locations. Disinfection is the primary goal, since aesthetic considerations such as taste, odour, appearance, and trace chemical contamination do not affect the short-term safety of drinking water.
- Solar water disinfection is a low-cost method of disinfecting water that can often be implemented with locally available materials. Unlike methods that rely on firewood, it has low impact on the environment.

ADDITIONAL TREATMENT OPTIONS

- *Water fluoridation*: In many areas fluoride is added to water for the purpose of preventing tooth decay. Fluoride is usually added after the disinfection process. In the U.S., fluoridation is usually accomplished by the addition of hexafluorosilicic acid, which decomposes in water, yielding fluoride ions.

- *Water conditioning*: This is a method of reducing the effects of hard water. Hardness salts are deposited in water systems subject to heating because the decomposition of bicarbonate ions creates carbonate ions that crystallise out of the saturated solution of calcium or magnesium carbonate. Water with high concentrations of hardness salts can be treated with soda ash which precipitates out the excess salts, through the common-ion effect, producing calcium carbonate of very high purity. The precipitated calcium carbonate is traditionally sold to the manufacturers of toothpaste. Several other methods of industrial and residential water treatment are claimed to include the use of magnetic or/and electrical fields reducing the effects of hard water.
- *Plumbosolvency reduction*: In areas with naturally acidic waters of low conductivity, the water may be capable of dissolving lead from any lead pipes that it is carried in. The addition of small quantities of phosphate ion and increasing the pH slightly both assist in greatly reducing plumbo-solvency by creating insoluble lead salts on the inner surfaces of the pipes.
- *Radium Removal*: Some groundwater sources contain radium, a radioactive chemical element. Typical sources include many groundwater sources north of the Illinois River in Illinois. Radium can be removed by ion exchange, or by water conditioning. The back flush or sludge that is produced is, however, a low-level radioactive waste.
- *Fluoride Removal*: Although fluoride is added to water in many areas, some areas of the world have excessive levels of natural fluoride in the source water. Excessive levels can be toxic or cause undesirable cosmetic effects such as staining of teeth. One method of reducing fluoride levels is through treatment with activated alumina.

OTHER WATER PURIFICATION TECHNIQUES

Other popular methods for purifying water, especially for local private supplies are listed below.

In some countries some of these methods are also used for large scale municipal supplies. Particularly important are distillation and reverse osmosis.

1. *Boiling*: Water is heated hot enough and long enough to inactivate or kill micro-organisms that normally live in water at room

temperature. Near sea level, a vigorous rolling boil for at least one minute is sufficient. At high altitudes three minutes is recommended. In areas where the water is "hard" , boiling decomposes the bicarbonate ions, resulting in partial precipitation as calcium carbonate. This is the "fur" that builds up on kettle elements, etc., in hard water areas. With the exception of calcium, boiling does not remove solutes of higher boiling point than water and in fact increases their concentration . Boiling does not leave a residual disinfectant in the water. Therefore, water that has been boiled and then stored for any length of time may have acquired new pathogens.

2. *Granular Activated Carbon filtering*: a form of activated carbon with a high surface area, adsorbs many compounds including many toxic compounds. Water passing through activated carbon is commonly used in municipal regions with organic contamination, taste or odors. Many household water filters and fish tanks use activated carbon filters to further purify the water. Household filters for drinking water sometimes contain silver to release silver ions which have an anti-bacterial effect.
3. Distillation involves boiling the water to produce water vapour. The vapour contacts a cool surface where it condenses as a liquid. Because the solutes are not normally vaporised, they remain in the boiling solution. Even distillation does not completely purify water, because of contaminants with similar boiling points and droplets of unvaporised liquid carried with the steam. However, 99.9% pure water can be obtained by distillation.
4. *Reverse osmosis*: Mechanical pressure is applied to an impure solution to force pure water through a semi-permeable membrane. Reverse osmosis is theoretically the most thorough method of large scale water purification available, although perfect semi-permeable membranes are difficult to create. Unless membranes are well-maintained, algae and other life forms can colonize the membranes.
5. The use of iron in removing arsenic from water.
6. *Direct contact membrane distillation*: Applicable to desalination. Heated seawater is passed along the surface of a hydrophobic polymer membrane. Evaporated water passes from the hot side through pores in the membrane into a stream of cold pure water on the other side. The difference in vapour pressure between the hot and cold side helps to push water molecules through.
7. *Gas hydrate crystals centrifuge method*: If carbon dioxide gas is mixed with contaminated water at high pressure and low temperature, gas hydrate crystals will contain only clean water. This is because

the water molecules bind to the gas molecules at molecule level. The contaminated water is in liquid form. A centrifuge may be used to separate the crystals and the concentrated contaminated water.

WATER PURIFICATION FOR HYDROGEN PRODUCTION

For the small scale production of hydrogen, water purifiers are installed to prevent formation of minerals on the surface of the electrodes and to remove organics and chlorine from utility water. First, the water passes through a 20 micrometre interference filter to remove sand and dust particles, then a charcoal filter using activated carbon to remove organics and chlorine and finally a de-ionizing filter to remove metallic ions. Testing can be done before and after the filter to verify the proper removal of barium, calcium, potassium, magnesium, sodium and silica. Another method that is used is reverse osmosis.

SAFETY AND CONTROVERSIES

Accidents have also been known to happen. In April, 2007, the water supply of Spencer, Massachusetts became contaminated with excess sodium hydroxide when its treatment equipment malfunctioned. Many municipalities have moved from free chlorine to chloramine as a disinfection agent. However, chloramine in some water systems, appears to be a corrosive agent. Chlormaine can dissolve the "protective" film inside older service line, with the leaching of lead into residential spigots. This can result in harmful exposure to lead, with elevated blood levels of lead the outcome. Lead is a known neurotoxin.

DEMINERALIZED WATER

Distillation removes all minerals from water, and the membrane methods of reverse osmosis and nanofiltration remove most to all minerals. This results in demineralized water which is not considered ideal drinking water. The World Health Organization has investigated the health effects of demineralized water since 1980. Experiments in humans found that demineralized water increased diuresis and the elimination of electrolytes, with decreased blood serum potassium concentration.

Magnesium, calcium, and other minerals in water can help to protect against nutritional deficiency. Demineralized water may also increase the risk from toxic metals because it more readily leaches materials from piping like lead and cadmium, which is prevented by dissolved minerals such as calcium and magnesium. Low-mineral water has been implicated in specific cases of lead poisoning in infants, when lead from pipes leached at especially high rates into the water. Recommendations for magnesium have been put at a minimum of 10 mg/L with 20–30 mg/L optimum; for calcium a 20 mg/L minimum and a 40–80 mg/L optimum, and a total water hardness of 2 to 4 mmol/L. At water hardness above 5 mmol/L, higher incidence of gallstones, kidney stones, urinary stones, arthrosis, and arthropathies have been observed. Additionally, desalination processes can increase the risk of bacterial contamination.

Manufacturers of home water distillers, of course, claim the opposite — that minerals in water are the cause of many diseases, and that most beneficial minerals come from food, not water. They quote the American Medical Association as saying "The body's need for minerals is largely met through foods, not drinking water."

The WHO report agrees that "drinking water, with some rare exceptions, is not the major source of essential elements for humans" and is "not the major source of our calcium and magnesium intake", yet states that demineralized water is harmful anyway.

"Additional evidence comes from animal experiments and clinical observations in several countries. Animals given zinc or magnesium dosed in their drinking water had a significantly higher concentration of these elements in the serum than animals given the same elements in much higher amounts with food and provided with low-mineral water to drink."

AGRICULTURAL WASTEWATER TREATMENT

Agricultural wastewater treatment relates to the treatment of wastewaters produced in the course of agricultural activities.

Agriculture is a highly intensified industry in many parts of the world, producing a range of wastewaters requiring a variety of treatment technologies and management practices.

NON-POINT SOURCE POLLUTION

Non-point source pollution from farms is caused by surface Run-off from fields during rain storms. Agricultural Run-off is a major source of pollution, in some cases the only source, in many watersheds.

Sediment Run-off

Soil washed off fields is the largest source of agricultural pollution in the United States.

Excess sediment causes high levels of turbidity in water bodies, which can inhibit growth of aquatic plants, clog fish gills and smother animal larvae. Farmers may utilize erosion controls to reduce Run-off flows and retain soil on their fields.

Common techniques include:

- Contour plowing
- Crop mulching
- Crop rotation

- Planting perennial crops
- Installing riparian buffers.

Nutrient Run-off

Nitrogen and phosphorus are key pollutants found in Run-off, and they are applied to farmland in several ways:

- Commercial fertilizer
- Animal manure
- Municipal or industrial wastewater or sludge.

These chemicals may also enter Run-off from crop residues, irrigation water, wildlife, and atmospheric deposition.

Farmers can develop and implement nutrient management plans to mitigate impacts on water quality:

- Map and document fields, crop types, soil types, water bodies
- Develop realistic crop yield projections
- Conduct soil tests and nutrient analyses of manures and/or sludges applied
- Identify other significant nutrient sources
- Evaluate significant field features such as highly erodible soils, subsurface drains, and shallow aquifers
- Apply fertilizers, manures, and/or sludges based on realistic yield goals and using precision agriculture techniques.

PESTICIDES

Pesticides are widely used by farmers to control plant pests and enhance production, but chemical pesticides can also cause water quality problems.

Pesticides may appear in surface water due to:

- Direct application
- Run-off during rain storms
- Aerial drift .

Some pesticides have also been detected in groundwater. Farmers may use Integrated Pest Management techniques to maintain control over pests, reduce reliance on chemical pesticides, and protect water quality. There are few safe ways of disposing of pesticide surpluses other than through containment in well managed landfills or by incineration. In some parts of the world, spraying on land is a permitted method of disposal.

POINT SOURCE POLLUTION

Farms with large livestock and poultry operations, such as factory farms, can be a major source of point source wastewater. In the United States, these facilities are called concentrated animal feeding operations or confined animal feeding operations and are being subject to increasing government regulation.

Animal Wastes

The constituents of animal wastewater typically contain:

- Strong organic content — much stronger than human sewage
- High solids concentration
- High nitrate and phosphorus content
- Antibiotics
- Synthetic hormones
- Often high concentrations of parasites and their eggs
- Spores of *Cryptosporidum* resistant to drinking water treatment processes
- Spores of *Giardia*
- Human pathogenic bacteria such as *Brucella* and *Salmonella*

Animal wastes from cattle can be produced as solid or semisolid manure or as a liquid slurry. The production of slurry is especially common in housed dairy cattle.

Treatment

Whilst solid manure heaps outdoors can give rise to polluting wastewaters from Run-off, this type of waste is usually relatively easy to treat by containment and/or covering of the heap. Animal slurries require special handling and are usually treated by containment in lagoons before disposal by spray or trickle application to grassland. Constructed wetlands are sometimes used to facilitate treatment of animal wastes, as are anaerobic lagoons. Excessive application or application to sodden land or insufficient land area can result in direct Run-off to watercourses, with the potential for causing severe pollution.

Application of slurries to land overlying aquifers can result in direct contamination or, more commonly, elevation of nitrogen levels as nitrite or nitrate. The disposal of any wastewater containing animal waste upstream of a drinking water intake can pose serious health problems to those drinking the water because of the highly resistant spores present in many animals that are capable of causing disabling in humans. This risk exists even for very low-level seepage via shallow surface drains or from rainfall run-off. Some animal slurries are treated by mixing with straws and composted at high temperature to produce a bacteriologically sterile and friable manure for soil improvement.

PIGGERY WASTE

Piggery waste is comparable to other animal wastes except that many piggery wastes contain elevated levels of copper that can be toxic in the natural environment. Ascarid worms and their eggs are also common and can infect humans if wastewater treatment is ineffective.

Treatment

As for general animal waste, although the liquid fraction of the waste is frequently separated off and re-used in the piggery to avoid the prohibitively expensive costs of disposing of a copper-rich liquor.

SILAGE LIQUOR

Fresh or wilted grass or other green crops can be made into the semi-fermented product called silage which can be stored and used as winter forage for cattle and sheep.

The production of silage often involves the use of an acid conditioner such as sulfuric acid or formic acid. The process of silage making frequently produces a yellow-brown strongly smelling liquid which is very rich in simple sugars, alcohol, short-chain organic acids and silage conditioner.

This liquor is one of the most polluting organic substances known. The volume of silage liquor produced is generally in proportion to the moisture content of the ensiled material.

Treatment

Silage liquor is best treated through prevention by wilting crops well before silage making. Any silage liquor that is produced can be used as part of the food for pigs. The most effective treatment is by containment in a slurry lagoon and by subsequent spreading on land following substantial dilution with slurry. Containment of silage liquor on its own can cause structural problems in concrete pits because of the acidic nature of silage liquor.

MILKING PARLOUR (DAIRY FARMING) WASTES

Although milk has a deserved reputation as an important and valuable food product, its presence in wastewaters is highly polluting because of its organic strength, which can lead to very rapid de-oxygenation of receiving waters.

Milking parlour wastes also contain large volumes of wash-down water, some animal waste together with cleaning and disinfection chemicals.

Treatment

Milking parlour wastes are often treated in admixture with human sewage in a local sewage treatment plant.

This ensures that disinfectants and cleaning agents are sufficiently diluted and amenable to treatment. Running milking wastewaters into a farm slurry lagoon is a possible option although this tends to consume lagoon capacity very quickly. Land spreading is also a treatment option.

SLAUGHTERING WASTE

Wastewater from slaughtering activities is similar to milking parlour waste although considerably stronger in its organic composition and therefore potentially much more polluting.

Treatment

Milking parlour wastes are often treated in admixture with human sewage in a local sewage treatment plant. This ensures that disinfectants and cleaning agents are sufficiently diluted and amenable to treatment. Running milking wastewaters into a farm slurry lagoon is a possible option although this tends to consume lagoon capacity very quickly. Land spreading is also a treatment option.

VEGETABLE WASHING WATER

Washing of vegetables produces large volumes of water contaminated by soil and vegetable pieces. Low levels of pesticides used to treat the vegetables may also be present together with moderate levels of disinfectants such as chlorine.

Treatment

Most vegetable washing waters are extensively recycled with the solids removed by settlement and filtration. The recovered soil can be returned to the land.

FIREWATER

Although few farms plan for fires, fires are nevertheless more common on farms than on many other industrial premises. Stores of pesticides, herbicides, fuel oil for farm machinery and fertilizers can all help promote fire and can all be present in environmentally lethal quantities in firewater from fire fighting at farms.

Treatment

All farm environmental management plans should allow for containment of substantial quantities of firewater and for its subsequent recovery and disposal by specialist disposal companies. The concentration and mixture of contaminants in firewater make them unsuited to any treatment method available on the farm. Even land spreading has produced severe taste and odour problems for downstream water supply companies in the past.

INDUSTRIAL WASTEWATER TREATMENT

Industrial wastewater treatment covers the mechanisms and processes used to treat waters that have been contaminated in some way by anthropogenic industrial or commercial activities prior to its release into the environment or its re-use. Most industries produce some wet waste although recent trends in the developed world have been to minimise such production or recycle such waste within the production process. However, many industries remain dependent on processes that produce wastewaters.

SOURCES OF INDUSTRIAL WASTEWATER

Iron and Steel Industry

The production of iron from its ores involves powerful reduction reactions in blast furnaces. Cooling waters are inevitably contaminated with products especially ammonia and cyanide. Production of coke from coal in coking plants also requires water cooling and the use of water in by-products separation.

Contamination of waste streams includes gasification products such as benzene, naphthalene, anthracene, cyanide, ammonia, phenols, cresols together with a range of more complex organic compounds known collectively as polycyclic aromatic hydrocarbons . The conversion of iron or steel into sheet, wire or rods requires hot and cold mechanical transformation stages frequently employing water as a lubricant and coolant.

Contaminants include hydraulic oils, tallow and particulate solids. Final treatment of iron and steel products before onward sale into manufacturing includes *pickling* in strong mineral acid to remove rust and prepare the surface for tin or chromium plating or for other surface treatments such as galvanisation or painting.

The two acids commonly used are hydrochloric acid and sulfuric acid. Wastewaters include acidic rinse waters together with waste acid. Although many plants operate acid recovery plants, where the mineral acid is boiled away from the iron salts, there remains a large volume of highly acid ferrous sulfate or ferrous chloride to be disposed of. Many steel industry wastewaters are contaminated by hydraulic oil also known as soluble oil.

MINES AND QUARRIES

The principal waste-waters associated with mines and quarries are slurries of rock particles in water. These arise from rainfall washing exposed surfaces and haul roads and also from rock washing and grading processes.

Volumes of water can be very high, especially rainfall related arisings on large sites. Some specialized separation operations, such as coal washing to separate coal from native rock using density gradients, can produce wastewater contaminated by fine particulate haematite and surfactants.

Oils and hydraulic oils are also common contaminants. Wastewater from metal mines and ore recovery plants are inevitably contaminated by the minerals present in the native rock formations. Following crushing and extraction of the desirable materials, undesirable materials may become contaminated in the wastewater.

For metal mines, this can include unwanted metals such as zinc and other materials such as arsenic. Extraction of high value metals such as gold and silver may generate slimes containing very fine particles in where physical removal of contaminants becomes particularly difficult.

FOOD INDUSTRY

Wastewater generated from agricultural and food operations has distinctive characteristics that set it apart from common municipal wastewater managed by public or private wastewater treatment plants throughout the world: it is biodegradable and Non-toxic, but that has high concentrations of biochemical oxygen demand and suspended solids . The constituents of food and agriculture wastewater are often complex to predict due to the differences in BOD and pH in effluents from vegetable, fruit, and meat products and due to the seasonal nature of food processing and postharvesting. Processing of food from raw materials requires large volumes of high grade water. Vegetable washing generates waters with high loads of particulate matter and some dissolved organics. It may also contain surfactants.

Animal slaughter and processing produces very strong organic waste from body fluids, such as blood, and gut contents. This wastewater is frequently contaminated by significant levels of antibiotics and growth hormones from the animals and by a variety of pesticides used to control external parasites.

Insecticide residues in fleeces is a particular problem in treating waters generated in wool processing. Processing food for sale produces wastes generated from cooking which are often rich in plant organic material and may also contain salt, flavourings, colouring material and acids or alkali. Very significant quantities of oil or fats may also be present.

COMPLEX ORGANIC CHEMICALS INDUSTRY

A range of industries manufacture or use complex organic chemicals. These include pesticides, pharmaceuticals, paints and dyes, petro-chemicals, detergents, plastics, paper pollution, etc. Waste waters can be contaminated by feed-stock materials, by-products, product material in soluble or particulate form, washing and cleaning agents, solvents and added value products such as plasticisers.

Nuclear Industry

The waste production from the nuclear and radio-chemicals industry is dealt with at *Radioactive waste.*

Water Treatment

Water treatment for the production of drinking water is dealt with elsewhere. Many industries have a need to treat water to obtain very high quality water for demanding purposes. Water treatment produces organic and mineral sludges from filtration and sedimentation. Ion exchange using natural or synthetic resins removes calcium, magnesium and carbonate ions from water, replacing them with hydrogen and hydroxyl ions. Regeneration of ion exchange columns with strong acids and alkalis produces a wastewater rich in hardness ions which are readily precipitated out, especially when in admixture with other wastewaters.

Treatment of Industrial Wastewater

The different types of contamination of wastewater require a variety of strategies to remove the contamination.

SOLIDS REMOVAL

Most solids can be removed using simple sedimentation techniques with the solids recovered as slurry or sludge. Very fine solids and solids with densities close to the density of water pose special problems. In such case filtration or ultrafiltration may be required. Although, flocculation may be used, using alum salts or the addition of polyelectrolytes.

OILS AND GREASE REMOVAL

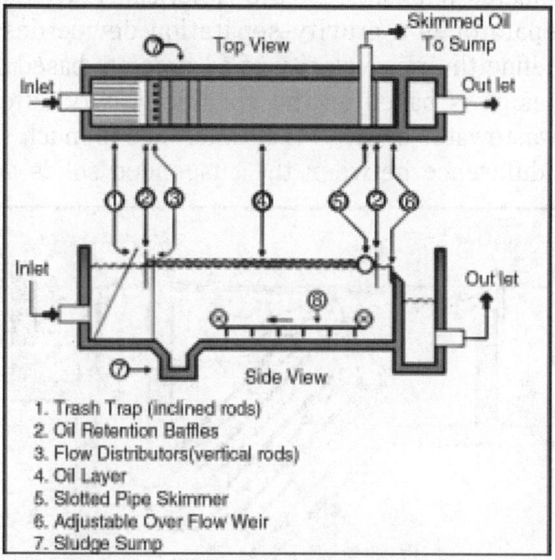

Fig. A Typical API Oil-water Separator Used in Many Industries.

Many oils can be recovered from open water surfaces by skimming devices. Considered a dependable and cheap way to remove oil, grease and other hydrocarbons from water, oil skimmers can sometimes achieve the desired level of water purity. At other times, skimming is also a cost-efficient method to remove most of the oil before using membrane filters and chemical processes.

Skimmers will prevent filters from blinding prematurely and keep chemical costs down because there is less oil to process.

Because grease skimming involves higher viscosity hydrocarbons, skimmers must be equipped with heaters powerful enough to keep grease fluid for discharge. If floating grease forms into solid clumps or mats, a spray bar, aerator or mechanical apparatus can be used to facilitate removal. However, hydraulic oils and the majority of oils that have degraded to any extent will also have a soluble or emulsified component that will require further treatment to eliminate.

Dissolving or emulsifying oil using surfactants or solvents usually exacerbates the problem rather than solving it, producing wastewater that is more difficult to treat.

The wastewaters from large-scale industries such as oil refineries, petrochemical plants, chemical plants, and natural gas processing plants commonly contain gross amounts of oil and suspended solids. Those industries use a device known as an API oil-water separator which is designed to separate the oil and suspended solids from their wastewater effluents.

The name is derived from the fact that such separators are designed according to standards published by the American Petroleum Institute.

The API separator is a gravity separation device designed by using Stokes Law to define the rise velocity of oil droplets based on their density and size. The design is based on the specific gravity difference between the oil and the wastewater because that difference is much smaller than the specific gravity difference between the suspended solids and water.

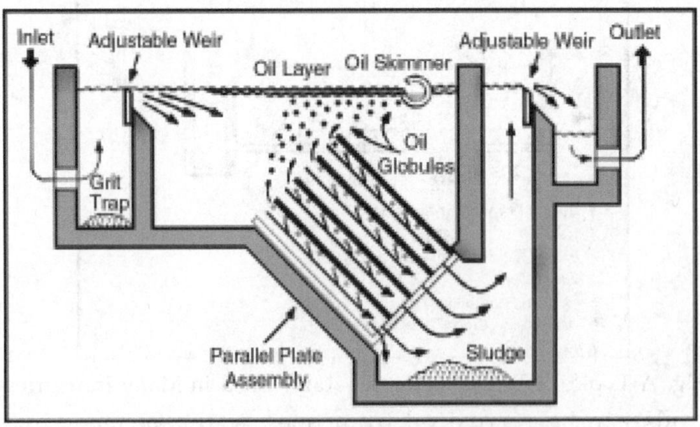

Fig. A Typical Parallel Plate Separator.

The suspended solids settles to the bottom of the separator as a sediment layer, the oil rises to top of the separator and the cleansed wastewater is the middle layer between the oil layer and the solids. Typically, the oil layer is skimmed off and subsequently re-processed or disposed of, and the bottom sediment layer is removed by a chain and flight scraper and a sludge pump. The water layer is sent to further treatment consisting usually of a Electroflotation module for additional removal of any residual oil and then to some type of biological treatment unit for removal of undesirable dissolved chemical compounds.

Parallel plate separators are similar to API separators but they include tilted parallel plate assemblies. The parallel plates provide more surface for suspended oil droplets to coalesce into larger globules. Such separators still depend upon the specific gravity between the suspended oil and the water.

However, the parallel plates enhance the degree of oil-water separation. The result is that a parallel plate separator requires significantly less space than a conventional API separator to achieve the same degree of separation.

REMOVAL OF BIODEGRADABLE ORGANICS

Biodegradable organic material of plant or animal origin is usually possible to treat using extended conventional wastewater treatment processes such as activated sludge or trickling filter. Problems can arise if the wastewater is excessively diluted with washing water or is highly concentrated such as neat blood or milk. The presence of cleaning agents, disinfectants, pesticides, or antibiotics can have detrimental impacts on treatment processes.

Activated Sludge Process

Activated sludge is a biochemical process for treating sewage and industrial wastewater that uses air and microorganisms to biologically oxidize organic pollutants, producing a waste sludge containing the oxidized material.

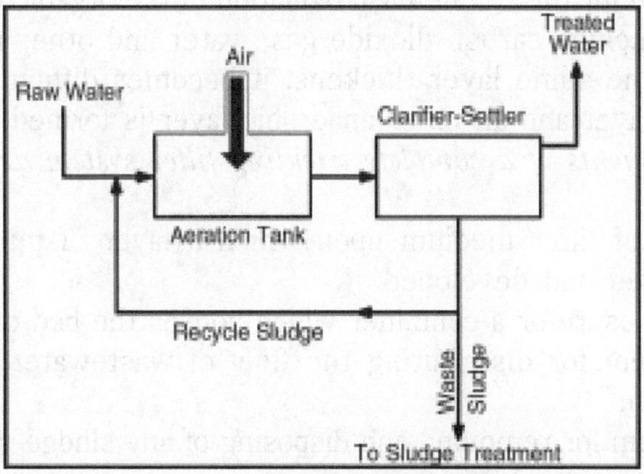

Fig. A Generalized, Schematic Diagram of an Activated Sludge Process.

In general, an activated sludge process includes:

- An aeration tank where air is injected and thoroughly mixed into the wastewater.
- A settling tank to allow the waste sludge to settle. Part of the waste sludge is recycled to the aeration tank and the remaining waste sludge is removed for further treatment and ultimate disposal.

Trickling Filter Process

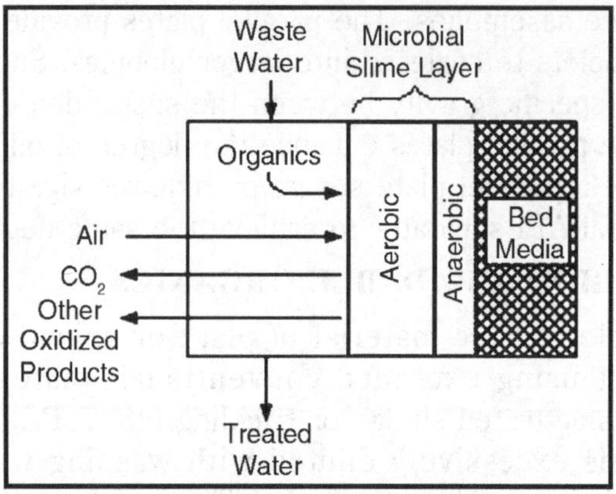

Fig. A Schematic Cross-section of the Contact Face of the Bed Media in a Trickling Filter.

A trickling filter consists of a bed of rocks, gravel, slag, peat moss, or plastic media over which wastewater flows downward and contacts a layer of microbial slime covering the bed media. Aerobic conditions are maintained by forced air flowing through the bed or by natural convection of air. The process involves adsorption of organic compounds in the wastewater by the microbial slime layer, diffusion of air into the slime layer to provide the oxygen required for the biochemical oxidation of the organic compounds.The end products include carbon dioxide gas, water and other products of the oxidation. As the slime layer thickens, it becomes difficult for the air to penetrate the layer and an inner anaerobic layer is formed.

The components of a complete trickling filter system are: fundamental components:

- A bed of filter medium upon which a layer of microbial slime is promoted and developed.
- An enclosure or a container which houses the bed of filter medium.
- A system for distributing the flow of wastewater over the filter medium.
- A system for removing and disposing of any sludge from the treated effluent.

The treatment of sewage or other wastewater with trickling filters is among the oldest and most well characterized treatment technologies. A trickling filter is also often called a trickle filter, trickling biofilter, biofilter, biological filter or biological trickling filter.

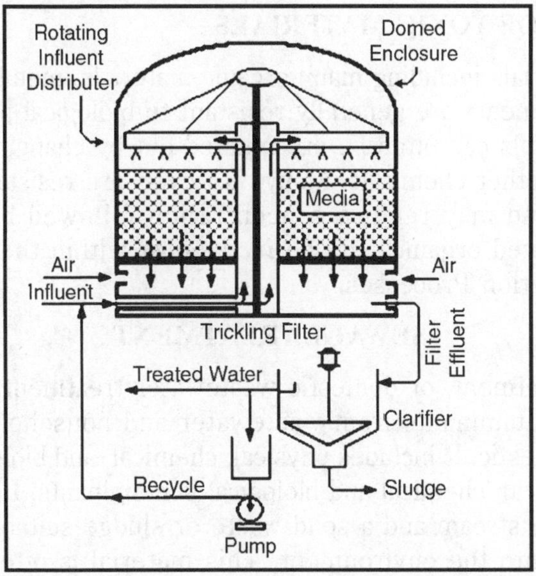

Fig. A Typical Complete Trickling Filter System.

TREATMENT OF OTHER ORGANICS

Synthetic organic materials including solvents, paints, pharmaceuticals, pesticides, coking products and so forth can be very difficult to treat. Treatment methods are often specific to the material being treated. Methods include Advanced Oxidation Processing, distillation, adsorption, vitrification, incineration, chemical immobilisation or landfill disposal. Some materials such as some detergents may be capable of biological degradation and in such cases, a modified form of wastewater treatment can be used.

TREATMENT OF ACIDS AND ALKALIS

Acids and alkalis can usually be neutralised under controlled conditions. Neutralisation frequently produces a precipitate that will require treatment as a solid residue that may also be toxic. In some cases, gasses may be evolved requiring treatment for the gas stream. Some other forms of treatment are usually required following neutralisation.

Waste streams rich in hardness ions as from de-ionisation processes can readily lose the hardness ions in a buildup of precipitated calcium and magnesium salts. This precipitation process can cause severe *furring* of pipes and can, in extreme cases, cause the blockage of disposal pipes. A 1 metre diameter industrial marine discharge pipe serving a major chemicals complex was blocked by such salts in the 1970s. Treatment is by concentration of de-ionisation waste waters and disposal to landfill or by careful pH management of the released wastewater.

TREATMENT OF TOXIC MATERIALS

Toxic materials including many organic materials, metals acids, alkalis, non-metallic elements are generally resistant to biological processes unless very dilute. Metals can often be precipitated out by changing the pH or by treatment with other chemicals. Many, however, are resistant to treatment or mitigation and may require concentration followed by landfilling or recycling. Disolved organics can be *incinerated* within the wastewater by Advanced Oxidation Processes.

SEWAGE TREATMENT

Sewage treatment, or domestic wastewater treatment, is the process of removing contaminants from wastewater and household sewage, both Run-off and domestic. It includes physical, chemical, and biological processes to remove physical, chemical and biological contaminants. Its objective is to produce a waste stream and a solid waste or sludge suitable for discharge or reuse back into the environment. This material is often inadvertently contaminated with many toxic organic and inorganic compounds.

ORIGINS OF SEWAGE

Sewage is created by residences, institutions, hospitals and commercial and industrial establishments. Raw influent includes household waste liquid from toilets, baths, showers, kitchens, sinks, and so forth that is disposed of via sewers. In many areas, sewage also includes liquid waste from industry and commerce.

The separation and draining of household waste into greywater and blackwater is becoming more common in the developed world, with greywater being permitted to be used for watering plants or recycled for flushing toilets. A lot of sewage also includes some surface water from roofs or hard-standing areas.

Municipal wastewater therefore includes residential, commercial, and industrial liquid waste discharges, and may include stormwater Run-off. Sewage systems capable of handling stormwater are known as combined systems or combined sewers. Such systems are usually avoided since they complicate and thereby reduce the efficiency of sewage treatment plants owing to their seasonality.

The variability in flow also leads to often larger than necessary, and subsequently more expensive, treatment facilities. In addition, heavy storms that contribute more flows than the treatment plant can handle may overwhelm the sewage treatment system, causing a spill or overflow. It is preferable to have a separate storm drain system for stormwater in areas that are developed with sewer systems.

As rainfall runs over the surface of roofs and the ground, it may pick up various contaminants including soil particles and other sediment, heavy metals, organic compounds, animal waste, and oil and grease. Some jurisdictions require stormwater to receive some level of treatment before being discharged directly into waterways. Examples of treatment processes used for stormwater include sedimentation basins, wetlands, buried concrete vaults with various kinds of filters, and vortex separators.

PROCESS OVERVIEW

Sewage can be treated close to where it is created or collected and transported via a network of pipes and pump stations to a municipal treatment plant. Sewage collection and treatment is typically subject to local, state and federal regulations and standards. Industrial sources of wastewater often require specialized treatment processes. Conventional sewage treatment involves three stages, called primary, secondary and tertiary treatment. First, the solids are separated from the wastewater stream. Then dissolved biological matter is progressively converted into a solid mass by using indigenous, water-borne micro-organisms.

Finally, the biological solids are neutralized then disposed of or re-used, and the treated water may be disinfected chemically or physically . The final effluent can be discharged into a stream, river, bay, lagoon or wetland, or it can be used for the irrigation of a golf course, green way or park. If it is sufficiently clean, it can also be used for groundwater recharge or agricultural purposes.

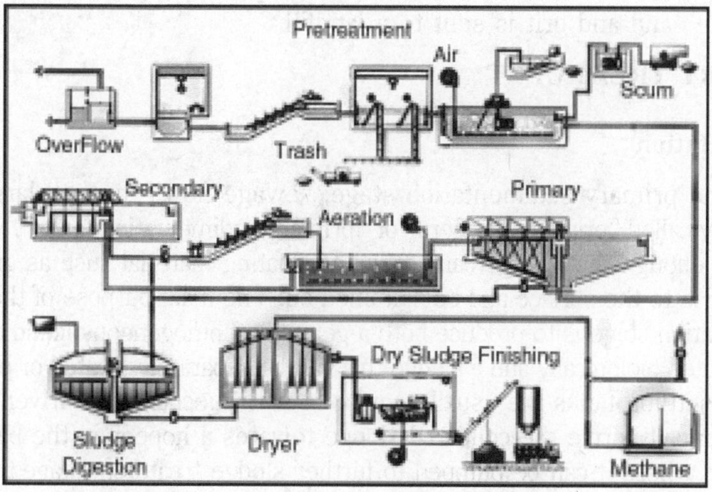

PRE-TREATMENT

Pre-treatment removes the materials that can be easily collected from the raw wastewater and disposed of. The typical materials that are removed during pre treatment include fats, oils, and greases , sand, gravels and rocks , larger settleable solids and floating materials. In modern plants serving large populations, sophisticated equipment with remote operation and control are employed whilst in smaller or less modern plants manually cleaned screen may be used.

SCREENING

The influent sewage water is strained to remove all large objects carried in the sewage stream, such as rags, sticks, tampons, cans, fruit, etc. This is most commonly done with a manual or automated mechanically raked bar screen. The raking action of a mechanical bar screen is typically paced according to the accumulation on the bar screens and/or flow rate. The bar screen is used because large solids can damage or clog the equipment used later in the sewage treatment plant. The large solids can also hinder the biological process. The solids are collected and later disposed in a landfill or incinerated. Pre treatment also typically includes a sand or grit channel or chamber where the velocity of the incoming wastewater is carefully controlled to allow sand grit and stones to settle, while keeping the majority of the suspended organic material in the water column. This equipment is called a de-gritter or sand catcher.

Sand, grit, and stones need to be removed early in the process to avoid damage to pumps and other equipment in the remaining treatment stages. Sometimes there is a sand washer followed by a conveyor that transports the sand to a container for disposal. The contents from the sand catcher may be fed into the incinerator in a sludge processing plant, but in many cases, the sand and grit is sent to a landfill.

PRIMARY TREATMENT

Sedimentation

In the primary sedimentation stage, sewage flows through large tanks, commonly called "primary clarifiers" or "primary sedimentation tanks". The tanks are large enough that sludge can settle and floating material such as grease and oils can rise to the surface and be skimmed off. The main purpose of the primary sedimentation stage is to produce both a generally homogeneous liquid capable of being treated biologically and a sludge that can be separately treated or processed. Primary settling tanks are usually equipped with mechanically driven scrapers that continually drive the collected sludge towards a hopper in the base of the tank from where it can be pumped to further sludge treatment stages.

Secondary Treatment

Secondary treatment is designed to substantially degrade the biological content of the sewage such as are derived from human waste, food waste, soaps and detergent. The majority of municipal plants treat the settled sewage liquor using aerobic biological processes. For this to be effective, the biota require both oxygen and a substrate on which to live. There are a number of ways in which this is done. In all these methods, the bacteria and protozoa consume biodegradable soluble organic contaminants and bind much of the less soluble fractions into floc.

Secondary treatment systems are classified as:

- Fixed-film or
- Suspended-growth.

Fixed-film treatment process including trickling filter and rotating biological contactors where the biomass grows on media and the sewage passes over its surface. In suspended-growth systems, such as activated sludge, the biomass is well mixed with the sewage and can be operated in a smaller space than fixed-film systems that treat the same amount of water. However, fixed-film systems are more able to cope with drastic changes in the amount of biological material and can provide higher removal rates for organic material and suspended solids than suspended growth systems. Roughing filters are intended to treat particularly strong or variable organic loads, typically industrial, to allow them to then be treated by conventional secondary treatment processes. Characteristics include typically tall, circular filters filled with open synthetic filter media to which wastewater is applied at a relatively high rate. They are designed to allow high hydraulic loading and a high flow-through of air. On larger installations, air is forced through the media using blowers. The resultant wastewater is usually within the normal range for conventional treatment processes.

SURFACE-AERATED BASINS

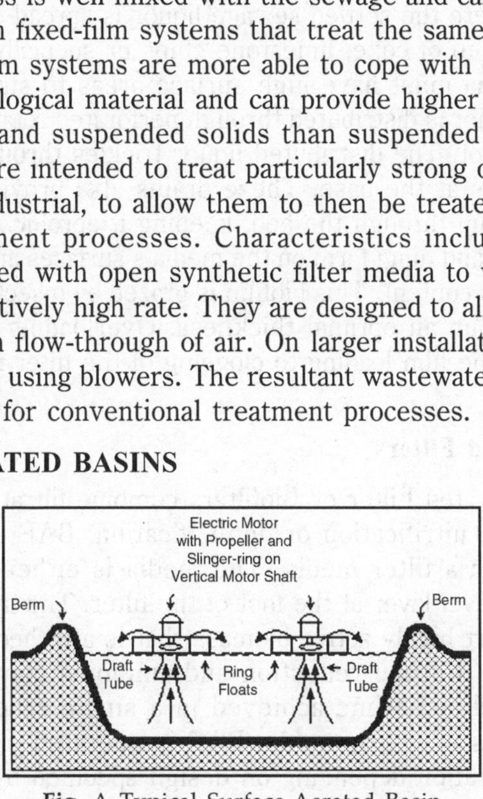

Fig. A Typical Surface-Aerated Basin

Most biological oxidation processes for treating industrial wastewaters have in common the use of oxygen and microbial action. Surface-aerated basins achieve

80 to 90% removal of Biochemical Oxygen Demand with retention times of 1 to 10 days. The basins may range in depth from 1.5 to 5.0 metres and use motor-driven aerators floating on the surface of the wastewater.In an aerated basin system, the aerators provide two functions: they transfer air into the basins required by the biological oxidation reactions, and they provide the mixing required for dispersing the air and for contacting the reactants. Typically, the floating surface aerators are rated to deliver the amount of air equivalent to 1.8 to 2.7 kg O_2/kW·h. However, they do not provide as good mixing as is normally achieved in activated sludge systems and therefore aerated basins do not achieve the same performance level as activated sludge units. Biological oxidation processes are sensitive to temperature and, between 0 °C and 40 °C, the rate of biological reactions increase with temperature. Most surface aerated vessels operate at between 4 °C and 32 °C.

Filter Beds

In older plants and plants receiving more variable loads, trickling filter beds are used where the settled sewage liquor is spread onto the surface of a deep bed made up of coke, limestone chips or specially fabricated plastic media. Such media must have high surface areas to support the biofilms that form. The liquor is distributed through perforated rotating arms radiating from a central pivot. The distributed liquor trickles through this bed and is collected in drains at the base. These drains also provide a source of air which percolates up through the bed, keeping it aerobic. Biological films of bacteria, protozoa and fungi form on the media's surfaces and eat or otherwise reduce the organic content. This biofilm is grazed by insect larvae and worms which help maintain an optimal thickness. Overloading of beds increases the thickness of the film leading to clogging of the filter media and ponding on the surface.

Biological Aerated Filters

Biological Aerated Filter or Biofilters combine filtration with biological carbon reduction, nitrification or denitrification. BAF usually includes a reactor filled with a filter media. The media is either in suspension or supported by a gravel layer at the foot of the filter. The dual purpose of this media is to support highly active biomass that is attached to it and to filter suspended solids. Carbon reduction and ammonia conversion occurs in aerobic mode and sometime achieved in a single reactor while nitrate conversion occurs in anoxic mode. BAF is operated either in upflow or downflow configuration depending on design specified by manufacturer.

Membrane Bioreactors

Membrane bioreactors combines activated sludge treatment with a membrane liquid-solid separation process. The membrane component uses low pressure microfiltration or ultra filtration membranes and eliminates the need for clarification and tertiary filtration. The membranes are typically immersed in the aeration tank. One of the key benefits of a membrane bioreactor system is that it effectively overcomes the limitations associated with poor settling of sludge in conventional activated sludge processes. The technology permits bioreactor operation with considerably higher mixed liquor suspended solids concentration than CAS systems, which are limited by sludge settling.

The process is typically operated at MLSS in the range of 8,000–12,000 mg/L, while CAS are operated in the range of 2,000–3,000 mg/L. The elevated biomass concentration in the membrane bioreactor process allows for very effective removal of both soluble and particulate biodegradable materials at higher loading rates. Thus increased Sludge Retention Times—usually exceeding 15 days—ensure complete nitrification even in extremely cold weather. The cost of building and operating a MBR is usually higher than conventional wastewater treatment, however, as the technology has become increasingly popular and has gained wider acceptance throughout the industry, the life-cycle costs have been steadily decreasing. The small footprint of MBR systems, and the high quality effluent produced, makes them particularly useful for water reuse applications.

Secondary Sedimentation

The final step in the secondary treatment stage is to settle out the biological floc or filter material and produce sewage water containing very low levels of organic material and suspended matter.

ROTATING BIOLOGICAL CONTACTORS

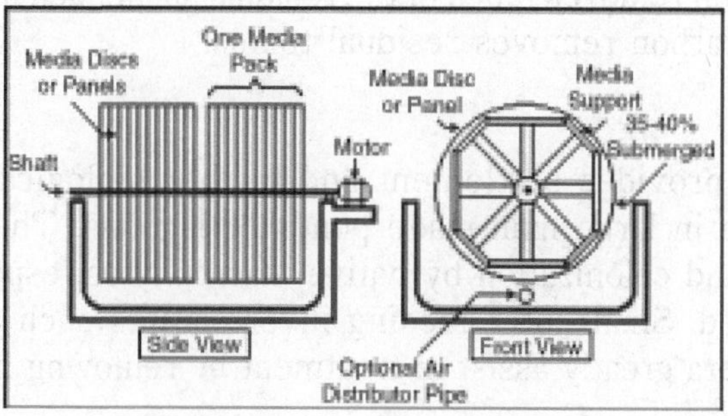

Fig. Schematic Diagram of a Typical Rotating Biological Contactor.

Rotating biological contactors are mechanical secondary treatment systems, which are robust and capable of withstanding surges in organic load. RBCs were first installed in Germany in 1960 and have since been developed and refined into a reliable operating unit. The rotating disks support the growth of bacteria and micro-organisms present in the sewage, which breakdown and stabilise organic pollutants. To be successful, micro-organisms need both oxygen to live and food to grow. Oxygen is obtained from the atmosphere as the disks rotate. As the micro-organisms grow, they build up on the media until they are sloughed off due to shear forces provided by the rotating discs in the sewage. Effluent from the RBC is then passed through final clarifiers where the micro-organisms in suspension settle as a sludge. The sludge is withdrawn from the clarifier for further treatment.

A functionally similar biological filtering system has become popular as part of home aquarium filtration and purification. The aquarium water is drawn up out of the tank and then cascaded over a freely spinning corrugated Fibre-mesh wheel before passing through a media filter and back into the aquarium. The spinning mesh wheel develops a biofilm coating of microorganisms that feed on the suspended wastes in the aquarium water and are also exposed to the atmosphere as the wheel rotates. This is especially good at removing waste urea and ammonia urinated into the aquarium water by the fish and other animals.

TERTIARY TREATMENT

The purpose of tertiary treatment is to provide a final treatment stage to raise the effluent quality before it is discharged to the receiving environment. More than one tertiary treatment process may be used at any treatment plant. If disinfection is practiced, it is always the final process. It is also called "effluent polishing".

Filtration

Sand filtration removes much of the residual suspended matter. Filtration over activated carbon removes residual toxins.

Lagooning

Lagooning provides settlement and further biological improvement through storage in large man-made ponds or lagoons. These lagoons are highly aerobic and colonization by native macrophytes, especially reeds, is often encouraged. Small filter feeding invertebrates such as Daphnia and species of Rotifera greatly assist in treatment by removing fine particulates.

CONSTRUCTED WETLANDS

Constructed wetlands include engineered reedbeds and a range of similar methodologies, all of which provide a high degree of aerobic biological improvement and can often be used instead of secondary treatment for small communities. One example is a small reedbed used to clean the drainage from the elephants' enclosure at Chester Zoo in England.

NUTRIENT REMOVAL

Wastewater may contain high levels of the nutrients nitrogen and phosphorus. Excessive release to the environment can lead to a build up of nutrients, called eutrophication, which can in turn encourage the overgrowth of weeds, algae, and cyanobacteria. This may cause an algal bloom, a rapid growth in the population of algae. The algae numbers are unsustainable and eventually most of them die.

The decomposition of the algae by bacteria uses up so much of oxygen in the water that most or all of the animals die, which creates more organic matter for the bacteria to decompose. In addition to causing deoxygenation, some algal species produce toxins that contaminate drinking water supplies. Different treatment processes are required to remove nitrogen and phosphorus.

Nitrogen removal

The removal of nitrogen is effected through the biological oxidation of nitrogen from ammonia (nitrification) to nitrate, followed by denitrification, the reduction of nitrate to nitrogen gas. Nitrogen gas is released to the atmosphere and thus removed from the water. Nitrification itself is a two-step aerobic process, each step facilitated by a different type of bacteria. The oxidation of ammonia to nitrite is most often facilitated by Nitrosomonas spp. Nitrite oxidation to nitrate, though traditionally believed to be facilitated by Nitrobacter spp. is now known to be facilitated in the environment almost exclusively by Nitrospira spp.

Denitrification requires anoxic conditions to encourage the appropriate biological communities to form. It is facilitated by a wide diversity of bacteria. Sand filters, lagooning and reed beds can all be used to reduce nitrogen, but the activated sludge process can do the job the most easily. Since denitrification is the reduction of nitrate to dinitrogen gas, an electron donor is needed.

This can be, depending on the wastewater, organic matter, sulfide, or an added donor like methanol. Sometimes the conversion of toxic ammonia to nitrate alone is referred to as tertiary treatment.

Phosphorus Removal

Phosphorus removal is important as it is a limiting nutrient for algae growth in many fresh water systems . It is also particularly important for water reuse systems where high phosphorus concentrations may lead to fouling of downstream equipment such as reverse osmosis.

Phosphorus can be removed biologically in a process called enhanced biological phosphorus removal. In this process, specific bacteria, called polyphosphate accumulating organisms, are selectively enriched and accumulate large quantities of phosphorus within their cells . When the biomass enriched in these bacteria is separated from the treated water, these biosolids have a high fertilizer value.

Phosphorus removal can also be achieved by chemical precipitation, usually with salts of iron , aluminum, or lime. This may lead to excessive sludge productions as hydroxides precipitates and the added chemicals can be expensive. Despite this, chemical phosphorus removal requires significantly smaller equipment footprint than biological removal, is easier to operate and is often more reliable than biological phosphorus removal. Once removed, phosphorus, in the form of a phosphate rich sludge, may be land filled or, if in suitable condition, resold for use in fertilizer.

DISINFECTION

The purpose of disinfection in the treatment of wastewater is to substantially reduce the number of microorganisms in the water to be discharged back into the environment. The effectiveness of disinfection depends on the quality of the water being treated , the type of disinfection being used, the disinfectant dosage , and other environmental variables. Cloudy water will be treated less successfully since solid matter can shield organisms, especially from ultraviolet light or if contact times are low. Generally, short contact times, low doses and high flows all militate against effective disinfection.

Common methods of disinfection include ozone, chlorine, or ultraviolet light. Chloramine, which is used for drinking water, is not used in wastewater treatment because of its persistence. Chlorination remains the most common form of wastewater disinfection in North America due to its low cost and long-term history of effectiveness.

One disadvantage is that chlorination of residual organic material can generate chlorinated-organic compounds that may be carcinogenic or harmful to the environment. Residual chlorine or chloramines may also be capable of chlorinating organic material in the natural aquatic environment. Further, because residual chlorine is toxic to aquatic species, the treated effluent must also be chemically dechlorinated, adding to the complexity and cost of treatment.

Ultraviolet light can be used instead of chlorine, iodine, or other chemicals. Because no chemicals are used, the treated water has no adverse effect on organisms that later consume it, as may be the case with other methods. UV radiation causes damage to the genetic structure of bacteria, viruses, and other pathogens, making them incapable of reproduction.

The key disadvantages of UV disinfection are the need for frequent lamp maintenance and replacement and the need for a highly treated effluent to ensure that the target microorganisms are not shielded from the UV radiation . In the United Kingdom, light is becoming the most common means of disinfection because of the concerns about the impacts of chlorine in chlorinating residual organics in the wastewater and in chlorinating organics in the receiving water. Edmonton and Calgary, Alberta, Canada also use UV light for their effluent water disinfection. Ozone O_3 is generated by passing oxygen O_2 through a high voltage potential resulting in a third oxygen atom becoming attached and forming O_3.

Ozone is very unstable and reactive and oxidizes most organic material it comes in contact with, thereby destroying many pathogenic microorganisms. Ozone is considered to be safer than chlorine because, unlike chlorine which has to be stored on site, ozone is generated onsite as needed. Ozonation also produces fewer disinfection by-products than chlorination. A disadvantage of ozone disinfection is the high cost of the ozone generation equipment and the requirements for special operators.

Package Plants and Batch Reactors

In order to use less space, treat difficult waste, deal with intermittent flow or achieve higher environmental standards, a number of designs of hybrid treatment plants have been produced. Such plants often combine all or at least two stages of the three main treatment stages into one combined stage. In the UK, where a large number of sewage treatment plants serve small populations, package plants are a viable alternative to building discrete structures for each process stage.

One type of system that combines secondary treatment and settlement is the sequencing batch reactor. Typically, activated sludge is mixed with raw incoming sewage and mixed and aerated. The resultant mixture is then allowed to settle producing a high quality effluent. The settled sludge is run off and re-aerated before a proportion is returned to the headworks. SBR plants are now being deployed in many parts of the world including North Liberty, Iowa, and Llanasa, North Wales, Southwestern United States.

The disadvantage of such processes is that precise control of timing, mixing and aeration is required. This precision is usually achieved by computer controls linked to many sensors in the plant. Such a complex, fragile system is unsuited to places where such controls may be unreliable, or poorly maintained, or where the power supply may be intermittent. Package plants may be referred to as high charged or low charged. This refers to the way the biological load is processed.

In high charged systems, the biological stage is presented with a high organic load and the combined floc and organic material is then oxygenated for a few hours before being charged again with a new load. In the low charged system the biological stage contains a low organic load and is combined with flocculate for a relatively long time.

Sludge Treatment and Disposal

The sludges accumulated in a wastewater treatment process must be treated and disposed of in a safe and effective manner. The purpose of digestion is to reduce the amount of organic matter and the number of disease-causing microorganisms present in the solids. The most common treatment options include anaerobic digestion, aerobic digestion, and composting. Choice of a wastewater solid treatment method depends on the amount of solids generated and other site-specific conditions. However, in general, composting is most often applied to smaller-scale applications followed by aerobic digestion and then lastly anaerobic digestion for the larger-scale municipal applications.

ANAEROBIC DIGESTION

Anaerobic digestion is a bacterial process that is carried out in the absence of oxygen. The process can either be *thermophilic* digestion, in which sludge is fermented in tanks at a temperature of 55°C, or *mesophilic*, at a temperature of around 36°C. Though allowing shorter retention time, thermophilic digestion is more expensive in terms of energy consumption for heating the sludge. One major feature of anaerobic digestion is the production of biogas, which can be used in generators for electricity production and/or in boilers for heating purposes.

AEROBIC DIGESTION

Aerobic digestion is a bacterial process occurring in the presence of oxygen. Under aerobic conditions, bacteria rapidly consume organic matter and convert it into carbon dioxide. The operating costs are characteristically much greater for aerobic digestion because of the energy costs needed to add oxygen to the process.

SLUDGE DISPOSAL

When a liquid sludge is produced, further treatment may be required to make it suitable for final disposal. Typically, sludges are thickened to reduce the volumes transported off-site for disposal. There is no process which completely eliminates the need to dispose of biosolids. There is, however, an additional step some cities are taking to superheat the wastewater sludge and convert it into small pelletized granules that are high in nitrogen and other organic materials.

In New York City, for example, several sewage treatment plants have dewatering facilities that use large centrifuges along with the addition of chemicals such as polymer to further remove liquid from the sludge. The removed fluid, called centrate, is typically reintroduced into the wastewater process. The product which is left is called "cake" and that is picked up by companies which turn it into fertilizer pellets. This product is then sold to local farmers and turf farms as a soil amendment or fertilizer, reducing the amount of space required to dispose of sludge in landfills.

TREATMENT IN THE RECEIVING ENVIRONMENT

Many processes in a wastewater treatment plant are designed to mimic the natural treatment processes that occur in the environment, whether that environment is a natural water body or the ground. If not overloaded, bacteria in the environment will consume organic contaminants, although this will reduce the levels of oxygen in the water and may significantly change the overall ecology of the receiving water. Native bacterial populations feed on the organic contaminants, and the numbers of disease-causing microorganisms are reduced by natural environmental conditions such as predation or exposure to ultraviolet radiation. Consequently, in cases where the receiving environment provides a high level of dilution, a high degree of wastewater treatment may not be required.

However, recent evidence has demonstrated that very low levels of certain contaminants in wastewater, including hormones and synthetic materials such as phthalates that mimic hormones in their action, can have an unpredictable adverse impact on the natural biota and potentially on humans if the water is re-used for drinking water.

In the US and EU, uncontrolled discharges of wastewater to the environment are not permitted under law, and strict water quality requirements are to be met. A significant threat in the coming decades will be the increasing uncontrolled discharges of wastewater within rapidly developing countries.

SEWAGE TREATMENT IN DEVELOPING COUNTRIES

There are few reliable figures on the share of the wastewater collected in sewers that is being treated in the world. In many developing countries the bulk of domestic and industrial wastewater is discharged without any treatment or after primary treatment only. In Latin America about 15% of collected wastewater passes through treatment plants. In Venezuela, a below average country in South America with respect to wastewater treatment, 97 percent of the country's sewage is discharged raw into the environment.

In a relatively developed Middle Eastern country such as Iran, Tehran's majority of population has totally untreated sewage injected to the city's groundwater. Israel has also aggressively pursued the use of treated sewer water for irrigation. In 2008, agriculture in Israel consumed 500 million cubic metres of potable water and an equal amount of treated sewer water. The country plans to provide a further 200 million cubic metres of recycled sewer water and build more desalination plants to supply even more water.

chal Most of sub-Saharan Africa is without wastewater treatment. Water utilities in developing countries are chronically underfunded because of low water tariffs, the inexistence of sanitation tariffs in many cases, low billing efficiency and poor operational efficiency. In addition, wastewater treatment typically is the process within the utility that receives the least attention, partly because enforcement of environmental standards is poor. As a result of all these factors, operation and maintenance of many wastewater treatment plants is poor. This is evidenced by the frequent breakdown of equipment, shutdown of electrically operated equipment due to power outages or to reduce costs, and sedimentation due to lack of sludge removal.

Developing countries as diverse as Egypt, Algeria, China or Colombia have invested substantial sums in wastewater treatment without achieving a significant impact in terms of environmental improvement. Even if wastewater treatment plants are properly operating, it can be argued that the environmental impact is limited in cases where the assimilative capacity of the receiving waters is high, as it is often the case.

IN DEVELOPING COUNTRIES

As of 2006, waterborne diseases are estimated to cause 1.8 million deaths each year. These deaths are attributable to inadequate public sanitation systems and it is clear that proper sewerage need to be installed. Appropriate technology options in water treatment include both community-scale and household-scale point-of-use designs. Military surplus water treatment units like the ERDLator are still seen in developing countries. Newer military style Reverse Osmosis Water Purification Units are portable, self-contained water treatment plants are becoming more available for public use. In order for the decrease of waterborne diseases to have long term effects, water treatment programs implemented by research and development groups in developing countries must be sustainable by its own residents. This can ensure the efficiency of such programs after the departure of the research team as monitoring is difficult because of the remoteness of many locations.

Bibliography

A. K. Kolay: *Soil Fertility*, Atlantic Publication, Delhi, 2010.

ARCEIVALA: *Wastewater Treatment for Pollution Control and Reuse , Third Edition*, Tata McGraw-Hill, Delhi, 2006.

Chandra A. Madramootoo, William R. Johnston and Lyman S. Willardson: *Management of Agricultural Drainage Water Quality*, Daya Publication, Delhi, 2005.

Charles Ernest Millar: *Soil Fertility*, Biotech Books, Delhi, 2004.

Ehrenfried Pfeiffer and E B Balfour: *Soil Fertility Renewal and Preservation : Bio-Dynamic Framing and Gardening*, Asiatic Publication, Delhi, 2006.

G Byju: *Soil Chemistry : Basic Concepts*, New India Publication, Delhi, 2008.

Henze: *Wastewater Treatment: Biological and Chemical Processes, 3e*, Springer, New York, 1999.

Hinrich L. Bohn, Brian L. McNeal and George A. OConnor: *Soil Chemistry*, Wiley, New York, 2011.

J.F. Gonzalez: *Wastewater Treatment in the Fishery Industry*, Daya Publication, Delhi, 2005.

J.P.J. Van Veren: *Soil Fertility and Sewage*, Agrobios Publication, Delhi, 2002.

Madireddi V. Subba Rao: *Water Conservation Management And Analysis*, Readworthy Publication, Delhi, 2011.

N.K. Prasad: *Soil Fertility and Plant Nutrition*, IBDC Publishers, Delhi, 2013.

Norman W. Hudson: *Soil and Water Conservation in Semi-Arid Areas*, Scientific Publication, Jaipur, 2004.

P C Bandyopadhyay: *Soil Fertility Management for Sustainable Agriculture*, Gene Tech Books, Delhi, 2007.

Philppe Quevauviller, Oliver Thomas and Andre Van Der Beken: *Wastewater Quality Monitoring and Treatment*, Wiley India, Delhi, 2013.

Pietro Laureano: *Water Conservation Techniques in Traditional Human Settlements*, Copal Publishing Group, Delhi, 2013.

R Manivanan: *Water Quality Modeling : Rivers, Streams and Estuaries*, New India Publication, Delhi, 2008.

R.K. Kaleeswari, R. Rajeswari and J. Prabhaharan: *A Handbook of Soil Fertility*, Satish Serial Publishing House, Delhi, 2013.

Rajendra Prasad and James F. Power: *Soil Fertility Management for Sustainable Agriculture*, CRC (Lewis Publishers), Delhi, 2012.

Ratan Kumar Saha: *Soil and Water Quality Management for Sustainable Aquaculture*, Narendra Publication, Delhi, 2010.

Ravinder Kumar Kashyap: *Soil Fertility, Fertilizer and Integrated Nutrient Management*, Oxford Book Company, Delhi, 2009.

S C Panda: *Soil Water Conservation and Dry Farming*, Agrobios Publication, Delhi, 2007.

S.K. Saini and Subhash Chandra: *Water Conservation and Utilization in Agriculture Production*, Satish Serial Publishing House, Delhi, 2010.

S.N. Kaul, Lidia Szpyrkowics and Arvind Kumar: *Wastewater Treatment Technologies and Environment (2 Vols-Set)*, Daya Publication, Delhi, 2004.

Shivanand Tolanur: *Soil Fertility, Fertilizer and Integrated Nutrient Management*, International, Delhi, 2006.

Shobha Rastogi and Rajesh K. Yadav: *Water Quality and Water Resources*, Oxford Book Company, Delhi, 2011.

Sriniwas Puri: *Soil Chemistry*, Oxford Book, Delhi, 2010.

Sujata V. Dhutraj: *Soil and Soil Fertility*, Axis Publication, Delhi, 2012.

Sukumar Bandyopadhyay: *Water Quality Management for Coastal Aquaculture*, Daya Publication, Delhi, 2008.

T.V.S. Prasad: *Soil Chemistry : Nutrient and Water Management in Agricultural Soils*, Dominant Publication, Delhi, 2009.

V.V. Dhruva Narayana: *Soil and Water Conservation Research in India*, Indian Council of Agricultural Research, 2002.

William J Ryam: *Water Treatment and Purification Technology*, Agrobios Publication, Delhi, 2007.

Index